Any Body's Guess!

Quirky Quizzes About What Makes You Tick

Michael J. Rosen

with Ben Kassoy

and bonus, bogus, and bonkers material

by M. Sweeney Lawless

Andrews McMeel Publishing, LLC

Kansas City • Sydney • London

 For information, write Andrews McMeel Publishing, LLC, an Andrews McMeel Universal company, 1130 Walnut Street, Kansas City, Missouri 64106.

10 11 12 13 14 SDB 10 9 8 7 6 5 4 3 2 1

ISBN-13: 978-0-7407-8991-5
ISBN-10: 0-7407-8991-0

Library of Congress Control Number: 2009939470

www.andrewsmcmeel.com

Any Body's Guess!

Contents

Any Body's Guess!

?

REVIZED
YoRe BODY
109th Edition
with over 6000 more
corrections + typos
queezy
spotty
creakies
funny feelings
owies
bumpy
still hurts
achy

What Makes Us Tick, Sick, and Thick

Your body doesn't have to be a scary thing. You're natural. You're normal. You're an amazing machine! (Granted, your operating manual has been poorly translated.) So you have questions.

Okay then! We have answers . . . in the form of questions.

The wondrous mechanism *you* is a fine piece of workmanship that discipline, exercise, and a decent diet can turn into a Formula One race car that purrs like a cat, pounces like a tiger, and scarfs down gazelles and giraffes as if they were chicken nuggets.

In the pages that follow, we present age-old *and* modern befuddlements that will both wrack your brain and pull your leg (two of our favorite body parts), at least at first.

What makes us tick? Ever wondered how the parts of your body acquired their names (the scientific names, that is)? Whom should you kiss if you want to reduce the risk of getting into a car accident? What should you drink if you want to avoid cataract surgery?

Your body has more exotic locales than any country, so kick back for a little armchair travel.

What makes us sick? Does the idea of bedbugs gross you out? Do you know that lice don't think we're so nice? Are you terrified of sharks but fond of honeybees?

Whether you're squeamish or curious, other creatures aren't just creepy-crawlies—they've been our fellow passengers in life throughout the ages.

What makes us thick? Do you wonder about weight? Do you work hard to unwind? Do you worry about whether you worry too much?

Those extra pounds, those extra hours of overtime, and those extra stresses may be affecting you, your family, and the planet in some weird and surprising ways.

With this book of fascinating facts and flippant questions in hand, you can challenge your friends, captivate your carpool, and entertain yourself when the airline announces yet another update on your delayed flight's status.

Go! Fight! Whine!

A common problem that used to plague adults almost exclusively now afflicts teenagers, especially cheerleaders, linebackers, and divers **(that's diving-board divers, not insane people who strap on parachutes and jump out of planes or who tempt sharks by swimming to the ocean floor)**. What's the problem, and what's happening to our youth?

The Make-No-Bones-About-It Answer

Aside from the fact that cheerleaders, linebackers, and divers all wear butt-hugging clothing, athletes in these sports are part of an expanding community of young people who suffer back injuries.

Aside from the common cold, lower back pain is responsible for the greatest number of sick days in the United States.

For jocks, jockettes, and Rockettes (yes, dancers also feel the pain), these strains and spasms result most often from overuse. (Coaches: "Three-a-days" should refer to meals, not workouts.) These days, adolescent back pain also stems from excess body weight, which puts extra pressure on the spine and back muscles, not to mention the living room recliner busted during the marathon video game session last Saturday night (inactivity leaves muscles flabby and weak).

Other things that cause your back to ache:

- **Lugging around heavy backpacks** (yes, metrosexuals, messenger bags count).
- **Improper lifting** (weights, girlfriend's books, girlfriend).
- **Poor posture** (slouching is the traditional way for a teenager to express disaffection for the world at large, disdain for the mall, and, for girls, the desire to be shorter than her date).
- **Tight hamstrings and weak abdominal muscles** (turns out that video games, while building great thumb strength, don't do the same for the hammies and abs).

So what can you do? Stop browsing videos of police chases on YouTube, go for a jog, and try doing a few extra crunches instead of a few extra munches.

Although back pain is becoming increasingly common in young athletes and those with growth spurts between the ages of ten and fifteen, it's generally benign and short-lived. So, young people, enjoy your cuteness while this acuteness lasts. Getting old is not child's play.

Don't Lose Your Head Over This Question

According to fairly reliable historical sources, two renowned prisoners, Marie Antoinette (1755–1793) and Sir Thomas More (1478–1535), both experienced the same shocking thing the morning of their beheadings. **Even if there's no such fate in your future, should you be worried?**

The White-as-a-Sheet Answer

The Queen of France was very fond of her elaborate hairstyling that some say resembled the diorama of a rainforest.

On the execution scaffold, Thomas More supposedly moved his beard clear of the chopping block, proclaiming, "It hath done no treason."

Alas, although these unfortunate icons were particular about their hair, history has it that both heads went suddenly and shockingly gray before they rolled.

Some say the intense anxiety and stress over their impending execution grayed Antoinette and More. What? Reigning as queen during the French Revolution or disrespecting Henry VIII's new bride hadn't provided a sufficient dose of each?

As we age, hair pigment follicles (we call them melanocytes—you can call them whatever you'd like) produce less pigment. As our hair falls out (on average, forty-five to sixty hairs each day), new hairs, with less melanin, are the replacements. Gradually—or suddenly—our heads sport an entirely different sort of gray matter.

In addition, an enzyme with the nifty name *catalase* starts to peter out as the decades pass, and the less catalase there is to break down naturally-accumulating hydrogen peroxide, the more color fades away. (Remember your days as a lifeguard at the city pool: H_2O_2 bleaches hair.)

before
behead

A few unhealthy conditions can also cause premature graying, including vitamin B_{12} deficiency, thyroid disorder, smoking, or fear or stress. (Your definition of *premature* will change over time—trust us.) Although research is inconclusive, some studies associate stress hormones with free radicals (not to be confused with people worried gray by McCarthy-era fanatics); these cell-damaging molecules may contribute to graying over the years.

But what of the Strange Phenomenon of Antoinette and More (SPAM, for short)? The answer isn't black and white; it's more of a gray area. One explanation is that they both suffered from the rare autoimmune condition known as diffuse alopecia areata, which causes rapid loss of nongray hairs, but this condition generally affects fifteen- to twenty-nine-year-olds, and Antoinette and More were executed at thirty-seven and fifty-seven, respectively. Another theory for the sudden grays is that they simply took off their wigs; however, Sir Thomas probably never wore one because the craze for manwigs hadn't begun yet.

All this teaches us all an important lesson: When you're executed, you can miss all kinds of up-and-coming trends!

Feeling Lousy?

"I am a citizen of the world. I am a dreamer. I am fresh. I am *so* not going to. . . ." This quote is part of an ad. What is it for, and what is it that the speaker doesn't want to do?

The All-Dressed-Up Answer

Although you and I might search with a fine-toothed comb (literally), German evolutionary anthropologists used more sophisticated means to date the remains of human body lice. And they hit upon something startling: Humans will date anything—even a louse, an unevolved louse who won't say "please" and "thank you" and who refuses to split the check—and that is sad. Oh, and they also revealed that our species went without clothing for about a million more years than scientists had previously thought.

In fact, the human body louse only infests clothing, so when hirsute humans traded epidermal wear for thermal wear, the body louse did find a new host, but all the other parasites who preferred our furrier climate—ticks, fleas, and other blood-sucking lice—decamped in search of other mammals. Gradually, survival of the fittest began to favor smooth skin. In a word, hairlessness became hot, a superior trait that made for more mating, dominance in the gene pool, and, in a blink of evolutionary time, a depilatory industry that today hovers close to $100 million each year. In fact, the quote mentioned earlier is for Nair Pretty, a hair remover targeted at girls ten to fifteen years old who are "*so* not going to have stubs sticking out" of their legs.

Although aquatic mammals lost their hair in exchange for improved hydrodynamics (the only reason it's acceptable for male swimmers to shave their legs), humans remain the only hairless primates. (Note: Robin Williams's forearms are not exactly human.) Along with a decrease in certain infestations, the loss of body hair helped keep us cool during our stay in the cradle of civilization, the African savanna. Still others claim that humans began pulling out their own hair as ancient stress levels over God and Darwin rose.

What all can agree on: Remember to keep your cool on your next date with a louse who's trying to get into your clothes.

Ten Sure-Fire Ways to Remove Hair (Some Even Work!)

A popular way to evict body lice is to get rid of body hair. Our species has a history of removing hair from wherever it grows (not to mention trying to grow it where it does not, but that's another topic). Men and women alike have devised many means of depilation, some more painful than others:

Shaving: The most popular means of hair removal is still using a razor to reduce a full beard to a goatee or a goatee to a mustache. The same razor (disguised in a pink plastic case with flowers on it) also reduces legs and underarms to, well, nakeder legs and nakeder underarms.

Plucking: Roman soldiers did it to get rid of their beards; modern women and manscaping men get rid of stray gray hair, send the old unibrow back to its own quarter, and restore the old nudity to something more nubile.

Waxing: Technically, this is plucking on a large scale rather than a one-by-one tweezing. (It's a day's task for a woman in bikini season or a man on the pro wrestling circuit, and it's reportedly more painful than being bodyslammed onto the canvas.)

Sugaring: All right, busted again. To women, this is plucking without tweezers or thread, waxing without the wax. To men, this is what a person does to coffee.

Mechanical depilation: Machines of all shapes and sizes are sold to remove unsightly hair or leave your skin smooth, but when it comes right down to it, they usually pluck.

Chemical depilation: Foams, gels, lotions, or sprays are applied to dissolve hair. Using enough of these will cause you to mutate into a hairless superhero.

Electrolysis: Electricity-conducting needles are injected into the hair shaft to kill the cells that grow hair. Not to be confused with electro convulsive therapy, which stands your hair on end like the bride of Frankenstein.

Lasers: These are concentrated beams of light that kill the cells that grow hair. (Be careful next time you're in the auditorium of an overzealous lecturer with a laser.)

Scissors: This is the same time-honored method that worked out so well for Delilah. But not so well for Samson.

Don't Lick Every Man in the Joint

Licking causes problems for 10 million people in the United States. Licking! But for the low, low fee of $3,950 **(plus $1,000 shipping and handling)**, their licking problem can be solved. What's the problem, and what does the sufferer get for the money?

The Lick of Truth

Certainly, $4,950 could buy a lifetime supply of lip balm or hire a personal assistant for a few months to seal envelopes and lick the candy shells off Tootsie Pops so you can get right to the chocolate centers.

Ailurophobes (people who fear cats) are usually afraid of cat bites or scratches, but 10 million other people in this country fret about cat saliva. Okay, so fretting is not the same thing as fearing, but if you suffer from cat allergies, it still haunts you. It's not the hair or the dander that causes the trouble. It's a specific protein in cat saliva that causes you to sneeze, cough, wheeze, paw your eyes, and consider every cat you see the devil's feather duster. Cat hair is covered with this irritant simply because a cat's idea of a trip to the beauty parlor is moving to a sunny spot and plastering down its hair with saliva.

At last, hope has arrived for those with cat allergies who still insist on having a cat. (Why? Why? People who fear water don't build themselves a diving pool. People afraid of heights don't spend a fortune erecting a master bedroom above the tree line.) But Ailurophiles (people who love cats so much that they even know the word *ailurophile*) with cat allergies can now buy—or at least put their names on a waiting list for—a hypoallergenic cat that sells for around $3,950, plus $1,000 shipping.

Who's Afraid of the Big, Bad Tabby?

People with cat allergies might very well develop a case of elurophobia (another spelling of *ailurophobia*); however, even if they don't, these furry friends give phobic folk plenty of other things to fear:

Aichmophobia: Fear of needles or sharp objects. (Claws!)

Pathophobia: Fear of disease. (Cat scratch fever, for instance.)

Felinophobia: One more word for a fear of cats. (There are lots of words—and cats, too.)

Agliophobia: Fear of pain. (Those claws are sharp!)

Melanophobia: Fear of the color black. (As in a black cat that crosses your path.)

Podophobia: Fear of feet. (Cats have four little ones, some with black toe pads.)

Doraphobia: Fear of fur. (Also comes in black.)

Wiccaphobia: Fear of witches. (A black cat's best pal.)

Somniphobia: Fear of sleep. (Cats do it a lot.)

Aclurophobia: Fear of cats. (People must fear them a lot to have this many words that mean the same thing.)

Gymnophobia: Fear of nudity. (Cats rarely wear pants.)

Pnigerophobia: Fear of being smothered. (Ever hear the old wives' tale warning against letting cats sleep with babies?)

Microphobia: Fear of small things. (Some cats are small.)

Megalophobia: Fear of large things. (Some cats are not small.)

Panophobia: Fear of everything. (Cats are included in everything.)

Polyphobia: Fear of many things. (For instance, cats and tortillas.)

Alektorophobia: Fear of chickens. (Okay, it's not about cats, but it makes a nice change of pace.)

Hierophobia: Fear of sacred things. (The ancient Egyptians worshipped cats.)

Galeophobia: A fear of sharks—or cats.

Gatophobia: We'll give you three guesses.

A Pigment of Your Imagination

In the 1940s, three quarters of all Americans could do this in their sleep. **Today, just over a tenth of Americans do.** What do they do, and how does it color their world?

The We're-Not-in-Kansas-Anymore Answer

Dang nabbit, what could it be, other than describing one's grandfather as a fuddy duddy? Dancing the jitterbug or bebopping to Ellington? Strolling to the cinema for the latest Orson Welles film?

Close, but no Freudian cigar. In the 1940s, most people dreamed in black and white.

Studies show that people under twenty-five now almost never dream in black and white, and people over fifty-five do so a quarter of the time. The difference, scientists tell us, has everything to do with the fact that many baby boomers watched little or no color TV as children. Looking at a black-and-white world, even a tiny square one, results in monochromatic dreams for adults.

In 1942, nearly 71 percent of college students reported seeing color rarely or never in their dreams, compared with only 18 percent in 2001. By the mid-1950s, the first color televisions were widely available for the first time, but for most folks it was either drive the new car to the movies or stay home and watch one on the new TV. A new Chevrolet and a new Admiral television each cost about a thousand bucks.

It's difficult to track the themes of dreams: Five minutes after one ends, you've forgotten half of it; ten minutes later, you've forgotten 90 percent of its content. That said, the most common dreams are those about falling, flying, and failing exams. Another recurring theme is not being able to find a restroom. Surely that was a living nightmare for the estimated 1.2 million people searching for one of the estimated 5,000 porta-potties at President Obama's inauguration. In simple terms, each john had to serve 240 people. If that's not the stuff of nightmares, what is?

Dream Works

People do all sorts of things when they dream, including work. Match the names of the brainiac with his or her brainchild:

Beautiful Dreamer

a. Stephen King
b. Madam C. J. Walker
c. Hannibal
d. Harriet Tubman
e. John James
f. Otto Loewi
g. Karlheinz Stockhausen

Beautiful Dream

1. A dangerous ambush
2. The nerve of some frog
3. A formula to grow hair
4. Helicopters and violins
5. The nurse, the leech, the house on the hill
6. One crazy travel itinerary
7. 27, 28, 29, 30, 42, and 43

Turn the page for the **answers**.

Who Snoozed What

a, 5: Stephen King says ideas that later became novels (*Misery, It, Salem's Lot*) first were dreams, or what the rest of us call nightmares.

b, 3: Madam C. J. Walker dreamed the ingredients for "Madam Walker's Wonderful Hair Grower," the product that made her the first female self-made millionaire in the United States.

c, 1: Legend says Hannibal's idea to invade Italy over the Alps with elephants arrived in a dream.

d, 6: Harriet Tubman had a dream that soldiers waited to attack her, so she traveled a different route and evaded capture.

e, 7: John James of Norfolk, England, won £150,000 in the lottery with six numbers plucked from a dream.

f, 2: Otto Loewi, "the Father of Neuroscience," literally dreamed up the experiment that proved—with the help of an unsung frog—how nerves communicate.

g, 4: Avant-garde German composer Karlheinz Stockhausen dreamed of the sound of strings and four helicopters hovering around a concert hall. Then he wrote an opera in which the string quartet does just that.

Five Minutes of Your Time

What do many people decide not to do, some people do for a few hasty seconds, **but surgeons do for five whole minutes?**

Come Clean with the Answer

They scream at nurses. They stab people. (They're real cut-ups, aren't they?) But there's one thing surgeons take very seriously indeed: They wash their hands.

Doctors don't just run their hands under the tap. They use hot water, plenty of soap, and special brushes to scour their fingernails, hands, and forearms, all the way up to the elbow. In fact, doctors are so serious about washing their hands that they don't call it "washing your hands"; they call it "scrubbing," which we thought meant putting on those green uniforms. But don't trust us; we're not doctors.

Why do doctors scrub (and why should you)? Germs are everywhere (computer keyboards, wet laundry, vacuum cleaners, bedding, shower curtains), and they're everywhere else (phones, doorknobs, litter boxes, sponges, faucets). Each time you high-five, put 'er there, or hand over that five bucks you borrowed, while you're saying, "Hey, Pal," your hands are saying, "Here's a lottery ticket for today's super lousy flu."

Germs also get around via droplet spread in "say it, don't spray it" moments. Particles of disease linger in the air, to be inhaled by anyone within three feet—about the size of what most of us consider our personal space. (Note to self: Increase personal space.)

Poor hand hygiene can also spread food-borne illnesses such as salmonella (even if that sounds like a fish who has lost her glass slipper, the charm ends there) and even *Escherichia coli,* which affects up to 76 million Americans every year with a variety of awful symptoms.

How do you vanquish germs when some of them can withstand a dose of radiation 1,000 times stronger than we can? It's simple: Increase the size of your personal bubble and increase your soap bubbles. In other words, avoid close contact with others, cover your coughs, and take it from the surgeons: Scrub to conquer dirt, bacteria, and viruses. Scrub for five minutes.

Feeling Especially Mysophobic?

Stressed out by staphylococcus? Beleaguered by botulism? Okay, so you're appropriately afraid of germs. Get out your mild solution of bleach and smite these top six germ havens on a weekly basis.

Popular Germ Hangouts	Bacteria Per Square Inch
Toilet bowl	3,200,000
Kitchen drain	500,000
Kitchen sponge	134,000
Bathtub	120,000
Kitchen faucet	13,000
Bathroom faucet	6,000

Don't Worry, Be Danish

1. Denmark
2. Switzerland
3. Austria
4. Iceland
5. Bahamas
6. Finland
7. Sweden
8. Bhutan
9. Brunei
10. Canada
11. Ireland
12. Luxembourg
13. Costa Rica
14. Malta
15. Netherlands
16. Antigua and Barbuda
17. Malaysia
18. New Zealand
19. Norway
20. Seychelles

These are, in order, the highest-ranking countries in terms of what?

The Happy-Stay-Lucky Answer

You're probably thinking, "Maybe blondes *do* have more fun," but then you get to the Bahamas. And then Bhutan, Brunei—you're not even sure where they are, are you?

These are the 20 happiest nations in the world.

And although happiness doesn't depend on diamonds, roses, champagne, or sappy poems (according to the first worldwide study of subjective well-being, anyway), it's not money but rather health that makes for happiness.

The study surveyed more than 80,000 people in 178 countries; that's less than .000013 of the world's population, so there's still a margin for error. (Apparently, a dozen or so countries didn't want to know how happy their populations are, so they weren't included.) The research revealed that in the list of ingredients determining a person's level of satisfaction, health comes first, followed by wealth, and then by education.

Researchers were surprised that Asian countries, despite being thought of as having a strong sense of collective identity, sailed east of happiness, with China in the #82 position, Japan at #90, and India trailing in the #125 spot. Along with China and India, the most populated countries, such as Russia (#167), tend to produce as much happiness as a backyard swimming pool in Siberia. In comparison, Malta has only 403,532 people to keep happy, so how hard can that be?

And what of the crabby old US of A, where amber waves of complaint ripple across mall parking lots everywhere? We're #23, despite the fact that only 36 percent of Americans described themselves as thriving in November 2008 (compared with 50 percent who felt that way back in January 2008, when the economy was better).

So, come on, everybody, let's beat Malaysia and conquer the Swiss! Here's what we can each do to raise ourselves, if not our country, higher in the rankings.

- **Surround yourself with joyful people** and send happiness back around, sort of like a game of Telephone or Pass the Orange.
- **Place yourself either at the center of social networks** or at the center of a tropical island such as Antigua and Barbuda (#16) or the Seychelles (#20).
- **Don't worry.** Be happy—or relocate to Denmark (#1).
- **Practice gratitude.** At least, be grateful that you're not in Zimbabwe (#177) or Burundi (#178).

Feel-Good Practice

Exercise is known to give people a sense of well-being, but if you're stuck somewhere you'd rather not be, there are a few things you can practice that are gentler on the joints than jogging and can also make you feel better quickly.

Breathe deeply. Oxygen feeds your tissues and replenishes your body's reserves. Just focusing on breathing from your diaphragm (your belly—not your ribs—should push out with each inhalation) can calm you down and fill you with energy. Not breathing can cause death, which is #1 among the reasons for unhappiness.

Rebalance your body. Whether you're standing in a crowded subway, slouching in line at a security checkpoint, or gripping your steering wheel as you stew in traffic, you have the power to make yourself more comfortable. Relax your jaw, ease your forehead, and yawn to release the muscles in your unhappy frown. Un-hunch your shoulders. Settle your weight evenly (whether you're on your feet or on your butt). Your word for the day is *unclench*.

Stretch. At your desk or in the doctor's waiting room, why not reach up both hands as far as you can and extend your spine until you feel an inch taller? Bend slowly, carefully, and touch your toes. Push yourself up from your chair as if you're on the parallel bars (you don't need chalk or a pair of stretchy tights; they make for an awkward boardroom). Pull in your gut as far as you can, count to ten, and then exhale. Repeat until you feel virtuous. Wiggle your toes, wriggle your fingers, roll your head around, shrug your shoulders and let them fall, tip back at a comfortable angle, and *breathe*.

Let your brain escape. What would make the current situation seem better? Remembering that vacation where the two of you watched pelicans at the beach, sipping Coronas, as if in a commercial? A rainbow? A foot massage from an attractive masseuse? Just visualize your favorite retreat. No one will know you're concentrating on something other than your work. What also helps with brain escapes: the Coronas.

Leave frustration to other people. You don't have to curse or jostle, bitch or sigh. If other people want to grumble or growl, that's their burden. Aggravation isn't a part of your job. Your job is to imagine your happy place: bowls of chips with endless salsa. Or maybe an endless salsa with a boy named Chip. It's up to you, really.

Hip or Not?

In a group of 100 adults, only 10 or 15 can. **But in a group of 100 elementary-age kids, about 85 can.** Can what? And does this make us adults long to be kids again?

Now here's a suggestion . . .

Tonight's Mesmerizing Answer

They can fall under the spell of a hypnotist. (The fact that they can eat an entire bag of Halloween candy in an hour is also amazing and nearly as mesmerizing.) Whereas adults have learned to limit their thinking, children have not exiled their imaginations to make room for such important un-fun things as manners, money, and how they will digest all that candy.

Is it because children are weak-willed, impressionable creatures that they are more likely to fall victim to hypnosis? Hardly. People under hypnosis are in a trance, but that does not mean they have no free will or awareness; instead of sleeping, they are actually in a hyper-attentive state.

Hypnotism accesses the thoughts and feelings of your subconscious, the uninhibited mind that acts first and lets you think later. This means the hypnotist, like the ideal spouse, can tap directly into your imagination, impulses, bodily sensations, and emotions—all functions that occur automatically and can be consciously directed. Because your conscious mind isn't filtering information during hypnosis, external suggestions seem to come directly from your subconscious, and that is why people in a trance perform outrageous things so willingly. (Has anyone hypothesized that rodeo clowns, stuntmen, and those *Jackass* guys are living in a hypnotized state?)

Children experience greater suggestibility than adults and relax into a more heightened imaginative state. Therefore, kids are more likely to say, "Hyp hyp hooray!" and respond more strongly to suggestions—except, of course, to ones about how it's time for bed ("You're getting sleepy, aren't you?") or about saving at least some of that Halloween candy until tomorrow.

Who Watches the Watch, Man?

Why should children have all the fun? Be a kid again and hypnotize yourself:

1. **Figure out why you would like to go into a trance.** (Remember, it's a state of hyper-awareness; you'll be communicating with your unconscious.) Do you want to stick to a new diet? Quit smoking? Write a novel? Keep your goal in mind while you undergo hypnosis.
2. **Relax.** Create conditions to make yourself comfortable. Choose a cozy chair and a room without gadgets, harsh light, or distractions such as obnoxious television, obnoxious cicadas, or obnoxious children—unless you've hypnotized them first. Play soft music or enjoy the quiet as you close your eyes, take gentle deep breaths, and focus on your goal.
3. **Let all the other images and thoughts fall away as you imagine yourself successfully achieving your goal.** Drink fresh water and push away the cake; it's too heavy to eat. Crumble your cigarettes into dust as your lungs fill with healthy, pure air. Find a positive statement to repeat to yourself: "I'm too happy for cake," or, "I now see that only Audrey Hepburn looked attractive smoking."
4. **Once you are un-hypnotized and re-energized, slowly climb back to real life.** Count backwards if you like. It's a new world without cake or cigarettes. Feel good as you begin the new life that achieving your goal makes possible.
5. **Think of these steps like a shampoo:** Repeat as often as necessary to reach your goal.

Freckled History

What type of mutant includes 2 percent of Germans, 4 percent of people in the world, **and probably about 1 percent of Neanderthals**?

The Better-Red-Than-Extinct Answer

Bill Walton, a 6-foot 11-inch, carrot-topped NBA Hall-of-Famer with a soft touch around the hoop and great court vision to go with it, is one of these mutants: redheads.

Red hair is caused not by pregnant mothers getting sunburns on their bellies (that creates blonde hair, duh!) but by an alteration of the MC1R gene. (FYI, this has no relation to the R2D2 gene, which determines whether your child will attend *Star Wars* conventions as an adult.)

The trait that produces gingers (not garlic—you still have to beware of vampires) as the British and Americans who watch *South Park* call them, is recessive. So both parents need a mutated version of the gene for it to appear in their freckled offspring. Even so, as long as the parents carry the mutated gene, the trait can skip a generation and reappear in a later one, regardless of the parents' hair color.

Despite the fact that our melanin-deprived brethren have been around for 20,000 to 40,000 years, rumors continue to circulate claiming that redheads could become extinct as early as 2060. And although this myth has been debunked—recessive genes don't disappear unless every carrier dies out or fails to reproduce—some truths about redheads remain:

- Although they have thicker hair, **they have 10,000 fewer hairs on their heads** than the average human scalp.
- **They make up 13 percent of Scotland's population**—the highest proportion of any country.
- **They're more sensitive to UV light,** which makes them prone to sunburn.
- **They are more susceptible to pain,** often needing more anesthesia and painkillers. Guess that explains why Walton missed 59 percent of his pro games due to injury.

Famous Mutant Humans

- Before **Mark Twain** became the familiar snowy-haired writer of humorous Americana, he was a redheaded lad named Samuel Langhorne Clemens. Witness the graying effects of a pseudonym.
- Italian astronomer **Galileo Galilei** lived out the last years of his life under house arrest for having the audacity to believe the earth rotated around the sun. His theories were vindicated many, many years later, when it was way too late for him to care. But Galileo isn't the only redhead who's been burned when it came to the sun.
- Before he was *not* the first redheaded U.S. senator, **John Glenn** *was* the first redheaded astronaut in space. Yes, we sent one of our mutants into space in search of other mutants.
- **Vincent van Gogh**.
- Reclusive poet **Emily Dickinson** may have simply been self-conscious about her auburn hair. Or maybe she just wanted to be left the hell alone.

- **Lynette "Squeaky" Fromme**, the follower of Charles Manson who tried to assassinate President Gerald Ford. She may have been a redhead, but pretty is as pretty does. "Squeaky"?
- The Soviet Union's first leader, **Vladimir Ilyich Ulyanov** (aka Lenin), was a natural redhead. The Soviet flag is red—coincidence?
- **Napoleon Bonaparte** and **Thomas Jefferson** had a lot in common: their red hair, a disdain for the British, and a willingness to let the other one keep South Dakota in the Louisiana Purchase.
- **L. Ron Hubbard** began his career writing science fiction, invented "Dianetics," and founded a Hollywood institution, Scientology. Ever wonder why you didn't know he was a redhead? His in-house lawyers didn't leave a stone unturned.

Why So Sensitive?

For the next three months, you've committed to sniffing a mushroom pizza every day. What are you doing besides supporting your local pizza delivery guy?

Anything-but-Dulled Answer

If, after deeply inhaling the aromas of each slice, you finally take a bite, and repeat this for several minutes or even hours, you're trying to appreciate the full experience of tantric eating. Or you're on one of those diets that condones this sort of torture. Or—and here is the truly restorative answer—you're trying to heighten your sense of smell by exercising it.

As it turns out, conscious practice will build up your senses just as reps build up your muscles; indeed, the opposite sex may be as impressed by your washboard abs as your ability to say, "Wasn't your perfume featured in the Victoria's Secret catalogue last spring?" Yes, an observation like this will place you head and shoulders above the pack of men who subscribe to the Victoria's Secret catalogue for . . . not perfume samples.

To hone the nose's sensitivity, choose one particularly pleasant odor—for example, a mushroom pizza—and take a whiff each day for a few months.

If you'd like to sensitize your taste buds, heed the adage "I came, I chewed, I swallowed," but do it slowly. Carefully notice each.

To improve your abs, take it easy with the pizza.

To improve hearing, listen to music at a moderate volume, focusing on one particular instrument. This should also keep you from hearing "superfly honey buns" rather than "sugar pie honey bunch." (And, as if red wine offered no other temptation, the resveratrol in fermented grapes can significantly improve your hearing; however, "wine goggles" will not improve your eyesight.)

You can improve your eyesight by spending less time in front of bright screens (computers and TVs, for example) and more time eating foods high in B vitamins such as riboflavin and thiamine, which are, completely coincidentally, ideal names for celebrity babies.

To improve their sense of touch and control, golfers use gloves for the practice swings and take them off for actual rounds. Removing gloves before you swing is not recommended for boxers. Come to think of it, if you're trying to become more sensitive to touch, boxing is not recommended.

In short, regular stimulation improves your senses. Try swimming in cold water. Indulge in hot stone massages. Try sex for "boosted sensitivity" (and the other reasons, too).

Slab Happy

You and I have a dozen, yet none of them are spare. A pigeon has five—and no spares there, either. **But a cow has thirteen**, and some of them are spare. What are they?

The Spare-Me-Your-Pity Answer

A cow has twenty-six ribs (thirteen pairs), a pigeon has ten ribs (five pairs), and humans have twenty-four ribs (it's normal to have exactly twelve pairs, whether you're a man or a woman, no matter what good book you've been reading).

Our ribs come in three varieties, none of them particularly delicious.

The top seven sets of ribs are attached to your breastbone by cartilage. Although your ribs are bone, without cartilage you would be unable to expand your chest to take a deep breath or to fill out an evening gown you have no business wearing.

Below your true ribs are three sets of ribs known as false ribs because they are not directly connected to the sternum; instead, they join up with the cartilage of the ribs above them.

The bottom two sets are called floating ribs because they connect only to the vertebrae of your spine.

Everyone's heard of Adam's rib (and a few of us have seen Katharine Hepburn and Spencer Tracy in George Cukor's movie of that name), but few of us remember that people used to take the concept of woman's creation from a rib literally. Sixteenth-century Flemish doctor Andreas Vesalius sparked outrage when he wrote his illustrated anatomical tome *De humani corporis fabrica* and let it slip that men and women have the same number of ribs.

Scientists now think that having no ribs on our lower spine gave humans—and mammals in general—an evolutionary edge. The configuration of our spine and ribcage allowed us to develop distinctive reproductive features (such as giving birth via the pelvis) and to create more flexible bodies back when speed and agility really counted, since Earth was populated by big hungry beasties with potentially delicious ribs to barbecue.

You Can't Do It While Sleeping—Or When Your Eyes Are Open, Either

Sunlight, overeating, **plucking your eyebrows**, and white pepper all have the potential to make you say one thing. What would that be?

The Answer (to your Prayers)

Salud! Or maybe *à tes souhaits!* You're not bilingual? How about *Gesundheit!* That's right, all those odd stimuli can induce a sneeze. (And, according to the Talmud, a sneeze means that your prayers will be answered, so *G-d bless you!*)

One hundred nine times larger than our Earth, the sun can do just about anything it likes: burn fiercely for millennia, provide energy to sustain all life, or win an NBA championship—the sun can do that. (But the Phoenix Suns? Not so much.) The sun can even cause some people to sneeze.

In fact, about a quarter of the world's population is affected by autosomal-dominant compelling helio-ophthalmic outburst, a syndrome with a rather obvious acronym: ACHOO. Scientists say this hereditary condition occurs because of a crossing of pathways in the brain. That is, your brain mistakes your eyes' normal reflex response to light for the sneezing reflex. This same twinge may also trigger a sneeze while you're plucking your eyebrows. For drivers with ACHOO, suddenly emerging into bright sunlight from a dimly lit tunnel can cause sneezing and therefore a potential hazard. It's also another good reason to avoid peppering your food while eating behind the wheel. (Stop for a minute and think of other bodily reflexes you can be thankful your brain doesn't confuse.)

So if you're driving over the speed limit and you have the flu, you're two accidents waiting to happen, whereas if you have the flu and sneeze without covering your mouth, you're just two versions of the same disaster.

So stay out of the sun, don't pluck while you drive, and sneeze into a tissue or your sleeve so you don't pepper people. When you sternute (that's sneezing vocabulary for the Pulitzer Prize–bound), your germ-carrying pitches, the snotball and the spitball, travel at speeds that exceed even that of a major league pitcher's fastball.

The Blessed

Wherever in the world you find yourself, friends and strangers alike may wish many wonderful things for you when you sneeze, but different cultures have different ways to salute such high-velocity emissions.

Blessings and Good Wishes

France: À tes souhaits (may your wish come true)
Iceland: Guð hjálpi þér (may God help you)

Good Health

Germany: Gesundheit
Holland: Gezondheid
Iran: Afiyat bashe (I wish you good health)
Poland: Na zdrowie (to your health)
Sweden: Prosit (to your benefit)
Switzerland: Gsondheit or xundheit (whichever you like better; we're neutral)

Long Life

Turkey: Çok yaşa (I wish you long life)
Vietnam: Sống lâu

Nothing

That's right, in countries such as Australia and Japan, it's the sneezer who talks—and only to say "Excuse me."

There's a Song in My Heart

The American Heart Association estimates that 100,000 to 200,000 lives could be saved each year while people are singing this song. **Who's singing? What song? And why?**

The Achy-Breaky Answer

Think back to the summer you spent as a camp counselor. Remember how irresponsible you were? That was you dozing in the lifeguard's chair, texting friends back home (or whatever primitive communication device you had back then) instead of belaying on the ropes course, and pretending you never saw the mattresses set below the cabin windows for campers leaping down after lights out. Think about all the times you narrowly avoided having to perform CPR.

But you've matured since then, so if—heaven forbid—you come upon a person whose heart has stopped beating, this time you'll be the hero. Make sure someone calls 911, send for the defibrillator (don't worry about spelling it—you don't have that kind of time), and, most importantly, *start singing*. Your memory of CPR might be fuzzy three months—let alone twenty years—after training, but it can all come back if you sing "Stayin' Alive."

Granted, your off-key falsetto impression of the Bee Gees' 1977 hit can rouse the dead, but that's not why it's useful: It approximates the perfect pace for resuscitative chest compressions (100 beats per minute). Yes, whether you're a brother or whether you're a mother, this old number from *Saturday Night Fever* is the number one hit in CPR training courses.

Since its introduction in 1960, CPR has had profound results: The presence of someone who knows CPR saves or prolongs the life of 70 percent of the victims of cardiac or respiratory emergencies.

Just throw yourself into the song. Don't worry about what people think about your Barry Gibb imitation. But do resist the John Travolta dancing.

The Beat Goes On

We don't have to tell you that you should be *certified* in CPR before you attempt to *perform* CPR, do we? Okay, just checking.

Here are a few other songs to help you the keep the beat until the heart starts beating:

"Another One Bites the Dust" by Queen. (Best not sung out loud in the presence of the next of kin.)

Mixed emotions? How about The Beatles' **"Hello Goodbye"** or Boy George's **"Do You Really Want to Hurt Me?"**

"Bailamos" by Enrique Iglesias. (Not recommended if you can't count in Spanish.)

"Dancing Queen" by ABBA. (The only danger is that it exposes the person administering the CPR to the risk of having that song stuck in his or her head for days.)

"Stars and Stripes Forever" will get that patriotic blood flowing again, as will the Marines' Hymn ("From the HALLS of MONteZU-hu-MA, to the SHORES of TRIP-o-LIIIIIIIIII").

"Piano Man," by Billy Joel (the *Live at Yankee Stadium* version) is ideal for people who like their choruses to be the same as their verses.

"Quit Playing Games with My Heart," by the Backstreet Boys. Just hope the victim gets the joke and that they're able to wake up and say, "I'm not! This is serious!"

"Baby Don't Forget My Number," by Milli Vanilli. If you don't feel confident with your voice, just move your lips while someone else sings.

Taxi! Taxi!

As you grow older, a part of your body shrinks, but if you drive a taxi—**and drive it for a while**—you can recover some of what you've lost. What will it take to make you consider installing a fare meter in your car?

Dad, You're Driving Us Crazy!

Taxi drivers have a lot on their minds: They take note of maniacs and the block-by-block possibility of reaching 60 miles per hour before braking at the next intersection; they navigate complicated detours (your tax dollars at work); they remember all the ridiculous things their passengers say (for the screenplay they've been working on since graduating from Columbia).

Although cabbies are known to lose it every so often (it's those *other* maniacs on the road), the thing they lose less than the rest of us is their car keys and brainpower. Driving a taxi helps the hippocampus, the part of our brain that forms and stores memories and specializes in spatial orientation. It's more developed in veteran cab drivers than in newer drivers or other commuters.

The Greeks named it the hippocampus for its resemblance to a seahorse (*hippos* means horse, and *kampos* means sea monster). Yes, it does make you wonder why eyebrows aren't called *kámpia* ("caterpillar") and the medulla oblongata isn't *pontíki kinoýmenwn schedíwn poy prospatheí na analábei ton kósmo* ("cartoon mouse trying to take over the world"). But we digress.

Yes, the word may be ancient, but the hippocampus itself is in a constant state of renewal because it possesses neuroplasticity, the ability to change shape and grow. It shrinks as we age (less for physically fit people), and it grows as we learn a musical instrument (that's why rock stars have such big heads) or study (the hippocampi of medical students preparing for finals increase in size).

As you mature, your brain grows significantly, to the point where you eventually realize cotton candy is the grossest food ever. Neuroplasticity also occurs after a head injury (ugh, the aluminum bat missed the piñata) and every time you memorize a poem, a formula, or something more complicated than Yellow Cab's phone number: 444-4444.

So keep learning, keep growing, and keep teaching that old seahorse new tricks.

Humor Is the Best Medicine—Or Sometimes Antibiotics Are

You're finally called from the waiting room into the doctor's office to get a tetanus shot. **(Let's leave aside for the moment whatever safety practice you must have ignored that put you here.)** If you're like most patients, you'll feel less pain if you do what?

The Big Pink Bunny of an Answer

Should you minimize the pain with a few Jell-O shots in the car on the way to the clinic? Would you feel better if your doc turned out to be your favorite of the Three Stooges?

But let's be serious—this is hard science. According to research conducted at New Mexico's Health Science Center, children, teenagers, and even adults are likely to register less pain (fewer shouts of "Owie!," fewer grimaces, fewer blubbery outbursts and calling for their mothers) if they're watching cartoons as the injection is given.

In an ideal world, your doctor would just steal into your bedroom while you're watching a Saturday morning Bugs Bunny marathon. With your kids, of course. Unfortunately, you're too old for house calls.

There's another way to reduce pain that you'd think you were too old to fall for: Patients report less pain if the syringe is decorated with brightly colored designs of bunnies, kittens, ducklings, flowers, and butterflies. (The study did not mention puppies, probably because they make people feel happy no matter what's happening.)

Why would painfully cute pictures be less painful? Needle aversion creates anxiety, often severe anxiety, in many people. Silly decorations short-circuit a patient's expectations, along with the fear and distress from the perceived threat of the needle. Your brain's fight-or-flight instincts are derailed by happy thoughts—or nauseated thoughts, if cutesy bunnies and kittens have that effect on you.

Sounds Like Trouble

You wake up one morning and your uvula is bothering you.
What sort of trouble is that, and who should you see?

The "Say Ah-Hah!" Answer

Does your insurance cover a visit to the uvulologist? Does such a thing even exist? Whatever it is, those two *u*'s in its name are certainly unusual. (*Unusual!* It's contagious! That has three *u*'s!)

Okay, ululating aside, if you look in the mirror and say "aaaaah," you can see your uvula hanging like Barbie's punching bag at the back of your mouth. *Uvula* is Greek for "little grape." (They obviously had no word for "tiny piñata.")

So what does it do? It's your switching yard, your drain stopper, your throat's traffic cop. And a lot depends on that drop of flesh. Air down the food tube (your esophagus) leads to belching and the words "pardon me"; food down the air tube (your windpipe) leads to choking and the Heimlich maneuver; and food up the air tube (into your nasal cavity) after a funny joke leads to every third grader's dream as a comic: the milk-out-the-nose spray.

Some scientists think the uvula is a hanger-on, coeval with your ungrateful toenails that provide so little service to you considering all the hours you spend clipping them and buying them new tube socks. Other folks suggest that we would lose important sounds in speech such as consonants and vowels that rely on that vibrating blob (for instance, the thrilling French trill of the rolling *r* in *au revoir*). So long to Native American, African, and Middle Eastern words that have consonants formed at the back of the throat.

Without a uvula, much of the noise we make while snoring would be impossible because it's the uvula that flutters like a sputtering chainsaw.

The good news? Although it can swell a bit from dehydration, allergies, infections, or irritation (smoking, for instance), your uvula runs maintenance-free for life. Now thank your uvula and finish the glass of milk before something funny happens.

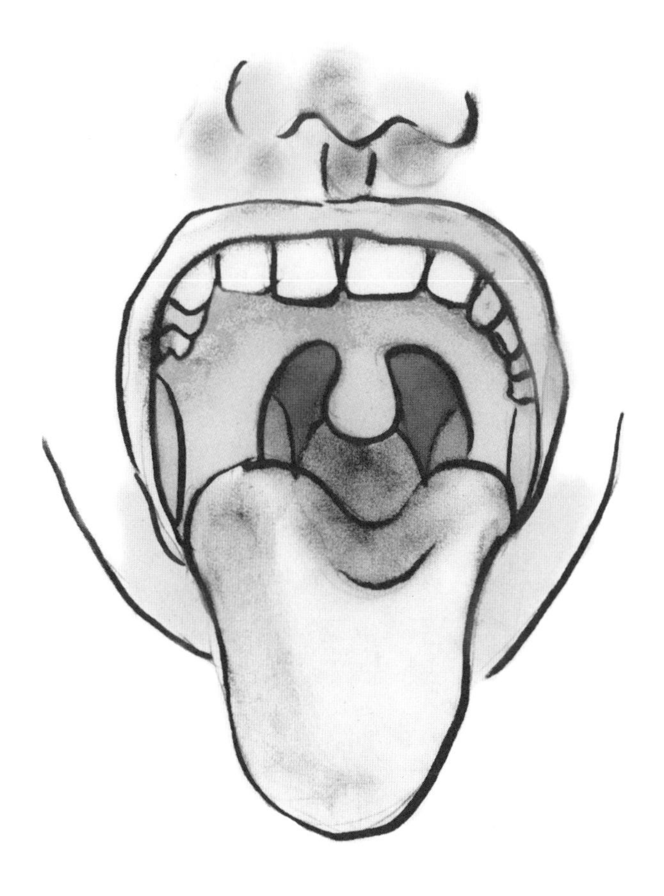

Me, me, me, meeeeeeeeee . . . but enough about me, how are you?

You Say Uvula, I Say Uvula

Like any hidden, damp body part, the uvula is rarely called by its medical name, especially in polite company. Below are slang terms (mostly made up by us) for the uvula that very well may have been popular at one point between forever ago and the future.

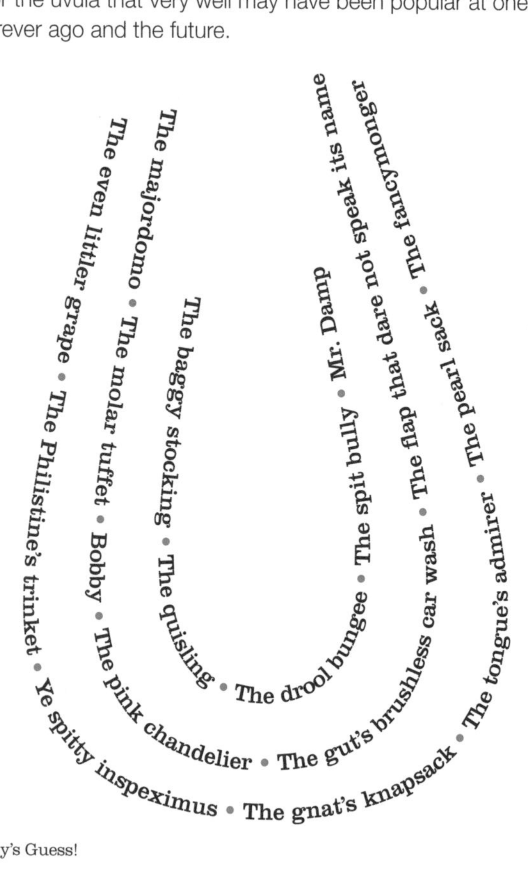

What Makes Perfect?

(The Answer Is Not "Steroids.")

It takes **10,000** what to become a world-class what?

The Very Disciplined Answer

Does it take 10,000 books, records, or manicures to become a world-class bibliophile, audiophile, or cosmetologist? Would 10,000 more leagues under the sea have made Nemo a better captain? Although you may wish otherwise—10,000 times over—it takes 10,000 hours of practice to become a world-class anything.

That's right, in order to be one of the world's best basketball players, chess players, or fencers, it takes about three hours of practice each day for a decade. Cognitive psychologists say that's the time necessary to achieve the required assimilation and consolidation of information in neural tissue. In other words, the more you practice, the more you remember and the less time you need to change gears. Until you put in the time needed to make your mental and physical processes fast, smooth, easy, and automatic, you'll keep riding the pine, losing your king to some computer, and getting stabbed in the face. Once you've improved your procedural memory, you're closer to mastering the ability to shoot the outside jump shot, win by checkmate with a zugzwang in the endgame, or stab someone in the face.

And while practice builds recall (and dominance in your chosen endeavor), it goes both ways. It's been proven that physical exercise decreases brain shrinkage as we age and decreases production of cortisol, a natural hormone that contributes to short-term memory loss and, in some cases, the development of Alzheimer's disease. In a new study where one group of subjects ran on a treadmill for a month while another remained sedentary, the exercise group developed 2.5 times more neurons, primarily in the hippocampus, the brain's center for learning and memory. The other group gained 2.5 times more knowledge of HBO, the TV's center for sex and gratuitous violence, but then promptly forgot everything they learned and couldn't remember what they'd done with the remote.

Not Just a Tall Tale

Fifty-eight percent of Fortune 500 company CEOs are this, **despite the fact that 85.5 percent of all U.S. men are not**, leaving them less fortunate in many senses.

The Tall and Short of It

Unfortunately for most men aspiring to become giants of industry, they got the short end of the deal: 58 percent of CEOs are more than 6 feet tall. Whereas the average CEO is just under 6 feet tall, the rest of the country's men fall short by 3 inches—not to mention several million dollars and a few very successful Ponzi schemes.

According to author and sociologist Malcolm Gladwell, tall people becoming rich people is no coincidence: They climb the corporate ladder faster with their longer legs because of our unconscious prejudice, which places more trust and likability in tall people. Gladwell also concluded that being shorter than average can be a disadvantage: White men who were 5 foot 6 inches and under were at just as much a disadvantage as African American and women CEOs. That is not to say that shorter men can't be successful; for instance, Kenneth Chenault is not only 5 foot 9 inches but one of less than 2 percent of Fortune 500 CEOs who are African American. However, height is a clear advantage for those who want to be head and shoulders above the rest.

One explanation for towering CEOs is that height, to a certain extent, is determined by a child's nutrition and exposure to diseases. Well-nourished, healthy children are able to reach their full physical and cognitive potential—whad'ya know, they're tall *and* smart!—which puts them in prime position to pursue higher education and prominent occupations. Tall people tend to be better educated than average. What's more, tall people are happier than other people as they look down on us and we look up to them.

On the other hand, the air isn't always better up there. At an average height of 4 feet 11 inches, Japanese women have a life span of about eighty-six years, the longest of any population in the world. Short people are also more likely to carry the so-called Methuselah gene, a genetic mutation that decreases the cell's use of insulin-like growth factor. The result is smaller bodies and a longer life—so there's a true David and Goliath story after all.

Figures to Figure for Your Figure—Go Figure!

How tall are you likely to be? Should you weigh less? And more importantly, can you afford that second slice of birthday cake? These mathematical formulas will keep you busy in the waiting room for your annual check-up.

Predict Your Height (Roughly)

1. Add together your mother's and father's height.
2. Divide the total by two.
3. If you're male, add 2½ inches. If you're female, subtract 2½ inches.

There's your projected adult height.

Find Your Ideal Weight

A quick and fairly accurate calculation is to give yourself 100 pounds if you're 5 feet tall. For each inch you stand above that, add 5 pounds. For instance, if you're 5 foot 10 inches that would be 100 pounds + 50 pounds = 150 pounds.

What's Your Ideal BMI?

Weight is *so* last millennium. It's not as crucial as your body mass index—you know, the ratio of your weight to your height. A healthy BMI is a number between 19 and 25.

Turn the page for the **formula**.

Here's the formula:

BMI = Weight in kilograms/Height in square meters

Here's how to use the formula:

1. Multiply your weight in pounds by 0.45 (this converts your pounds to kilograms).
2. Multiply your height in inches by 0.025 (this converts inches to meters).
3. Square your height. (That just means to multiply the number by itself.)
4. Finally, divide your weight by your squared height.

Now say, "Thank you, Pocket Calculator. I owe you one."

Which brings us to the really important issue: Should you have another piece of cake? If you're active, if your BMI is pretty good, and if you did eat all your veggies, sure, have a little slice.

Do Children Count?

If you're Chinese, you can reach forty at age four, but in the United States, at four years old you can only make it to around fifteen.
Exactly what kind of numbers racket is this?

The What's-My-Locker-Combination-Again Answer

If your toddler's first language is Chinese (Cantonese, that is—the language spoken on the mainland, in Hong Kong, and elsewhere) she is likely to be able to count higher than her European American cousins who speak English (the language spoken in England, New England, and elsewhere thanks to fast food and rap music).

Why can Cantonese-speaking children count higher than their English-speaking counterparts? Is it because Chinese culture puts such a premium on serious childhood pursuits such as study and academic achievement? Maybe, although few four-year-olds earn Ph.D.s (honorary doctorates don't count). The answer is simpler: Most Cantonese-speaking children can count higher because Cantonese numbers generally make more logical sense than English ones.

"A number is a number," you protest. "The whole thing just doesn't add up." Well, since you count in English, I'll go slowly. Try to act surprised when I tell you that English is highly irregular. Take the number fourteen: Instead of calling it "ten-and-four," we reverse the order of the digits, and the first syllable refers to the ones place and the second means the tens place. Although we follow suit with fiveteen (sort of), sixteen, seventeen, eighteen, and nineteen, two-and-ten is not "twoteen," and "eleven" just seems like a crazy typo. When we get into the twenties, we lose our minds and switch the ones and tens places so instead of three-and-twenty we pull a switcheroo and it's twenty-three.

Cantonese, on the other hand, keeps it simple, short, and logical: Twenty-three is "two-tens-and-three," and fourteen is "ten-and-four." (Twenty-four is the number of NBA star Kobe Bryant, the best-selling jersey in both countries.) Therefore, counting (and later, adding) is easier for Cantonese children, so it's no wonder a Chinese four-year-old can count to an average of forty, whereas Americans make it only to fifteen.

Hey, English-speaking toddlers, let's put those noses to the grindstone!

Sophomoric Problems

Remember your old friend algebra? Okay then! If you solve this equation, what will you discover?

$$(S + C) \times (B + F)/T = V$$

The Now-We've-Hit-Bottom Answer

Wait, don't turn the page yet! What if we told you that solving the equation solves the age-old problem of assessing who has the best butt? David Holmes, a psychologist at Manchester Metropolitan University, devised the formula and claims to have caused at least one football scholarship recipient to be "totally turned on to math for the first time in, like, ever."

The glute variables are as follows: *S* is "overall droopiness," *C* is "how spherical the buttocks are," *B* is the backside's "muscular wobble," *F* is for "firmness," *T* is for the amount of cellulite, and *V* represents the symmetry of the bottom. So you're more of a humanities person? Holmes sums up the ideal poetically: "The perfect female derriere has firmness to the touch and a resilience that prevents undue wobble or bounce, yet looks soft with flawless skin."

Astoundingly, he has produced no such formula for the male derriere.

The best way to obtain these particularly charming cheeks, suggests a University of Wisconsin study, is the quadruped hip extension. Start on hands and knees; bend one leg at a 90-degree angle and lift it until your foot is parallel to the ceiling and the upper leg aligns with the torso. Yes, you may look silly doing this, but you'd look even sillier paying a personal trainer $100 an hour to tell you to do the same thing.

The side effects of an evenly proportioned face offer even more reason to look on the bright side. Evolutionary psychologist Nick Neave states, "You're attracted to [a person with a symmetrical face] on a subconscious level because, throughout history, humans have chosen to breed with people they perceive to be healthy. Healthy genes mean a symmetrical face." More proof of survival of the facially fittest: Studies show that proportional features improve a person's ability to resist infection, which includes just about everything but lovesickness.

But getting back to the butt: There are flaws in the formula. For instance, our modern standards of beauty aren't the same as the Rubenesque ideal of the 1600s; they're not even standard across cultures and countries of the present day. So it's good to remain open to many forms of physical beauty; after all, eventually Olive Oyl and Popeye found each other.

Crikey, It's Cold!

(or "Gujirritj-yi ngan-ma-n!" in Wagiman)

In Australia, instead of using degrees Fahrenheit or Celsius, the Aborigines described how cold a night was in quite different terms. **Can you guess how?**

The Indigenous Insulation Indicator

How did they describe the cold? In words—probably Aborigine ones. Or maybe in barks, pants, and wet-nosed sniffles. According to nineteenth-century accounts, certain Aborigines rated the night's chilliness by how many dogs it took to keep themselves warm. (Apparently, some Australian Aborigines, who gave dingoes sleeping privileges in their huts, were not the first to ask the question, "Who let the dogs in?" Pre-Aztecs were said to use the Mexican hairless as nighttime hot water bottles.)

When it comes to temperature, humans have withstood the highest highs (262°F by a scientist in 1948) and the lowest lows. Serious hypothermia begins at just 3 degrees below normal body temperature, and below 90°F is an emergency, causing depressed breathing, falling heart rate, and low blood pressure.

So in terms of body temperature, average will do just fine; it just might not be the average to which you're accustomed. Carl Wunderlich, a nineteenth-century doctor, is credited with collecting the armpit temperatures of 25,000 patients (a job that would have been the pits, had the physician not taken so much delight in tickling others); he's the person responsible for the 98.6°F body temperature long held as the benchmark for normal. However, more recent studies have pegged average body temperature at about 98.1° or 98.2°F. It's possible Wunderlich can blame his foot-long mercury thermometer, which had to be held under a patient's pit for about eleven minutes; it could have been incorrectly calibrated.

The facts can now be seen in the cold light of day: Dr. Wunderlich should have sweated more over his results. For nearly 200 years, his findings caused heated debate. At last, we've warmed to the fact that our temperature is half a degree cooler than 98.6°F.

How cool is that?

Speeding Recovery

Your operation went forward without a hitch, but recuperation has taken longer than expected. **What about the drive to the hospital made this so?**

The Kiss-Will-Make-It-All-Better Answer

En route to the hospital for your surgery, there was squabbling: *Someone* was driving too fast and *someone* forgot to turn down the thermostat and *someone* forgot to remind the cat sitter about the doggie bag with half a crab cake in the fridge. Maybe stress about the surgery led to the complications in the car, but what if it also added complications to the surgery?

According to new research, arguments can slow the body's healing: Thirty minutes of squabbling can add another day or more, and constant marital mayhem can double recovery time. (Okay, that's not such a bad thing if you're hoping not to return to a hostile homestead.) But arguing increases your health risks overall.

In one study, forty-two couples had eight small wounds created on their arms during two 24-hour hospital stays. In the first visit, spouses were asked to talk about a personality trait each would like to change about themselves but which was not a source of tension in their relationship. In the second visit, they discussed a marital problem that aroused strong negative feelings. The results: Wounds after the second visit took longer to heal. And the wounds of high-hostility couples took even more time.

Here's more proof: Dental students who submitted to small oral wounds during finals week needed three extra days of healing compared with students whose wounds were administered during spring break.

When stress slows healing, one of the culprits is the production of cytokine, the cells' messenger protein, which summons immune cells to the site of your owie. Stress increases the amount of cytokines you produce, and chronic stress not only maintains too many in your system but also reduces your ability to produce enough where they're most needed.

In the midst of all the stress and stressing about stress, you might be interested to know that people who feel a deep sense of reverence and personal sense of spirituality in their lives have fewer postoperative complications and overall have speedier recoveries.

The bottom line: Love your life. Love your spouse. Have surgery on your honeymoon.

Can You Stomach It?

The Okinawans **(residents of the largest of Japan's Ryukyu Islands)** have a habit known as *hara hachi bu,* and many believe it's the reason they live longer and healthier lives than most people on Earth. Why don't people in the United States practice *hara hachi bu,* which is (a) a martial art, (b) a form of meditation, or (c) a dietary practice?

The Half-Full Answer

First, *hara hachi bu* is Japanese, so maybe the reason most of us don't do it is that we don't understand Japanese. But even if we did, we're not, as a culture, ready to embrace the concept of eating only until the stomach is 80 percent full; that's an idea reserved for a dinner of leftovers for the third night in a row or for surviving in wartime.

Indeed, Okinawans still dine as they did in wartime. Those raised before World War II, when fighting ravaged the Pacific archipelago, never experienced the bliss of overindulgence; in fact, the generation that lost one third of its population experienced terrible privation. So while the victors returned to their homes of suburban luxury and to their jars of pureed pears and instant oatmeal, the Okinawans continued to nosh sparingly on home-grown produce. A typical Okinawan meal of vegetables, tofu or miso soup,

and a small portion of fish or meat has fewer calories then a hamburger (and doesn't come with a toy). While some North Americans toy with the South Beach Diet, these Pacific islanders have had a lifetime of their own beach diet, never having shared in the modern prosperity enjoyed by much of Japan.

Their lack of modern conveniences (and the unhealthy Western diet that seems to accompany them) left Okinawans little choice but to practice the Confucian principle of *hara hachi bu*. The silver lining is that, for the most part, they enjoy a greater *ikigai* (sense of well-being) and a culture of *yuimaru* (a supportive community). Not only is the Okinawans' life worth living, but their life expectancy is among the highest in the world: seventy-eight years for men, eighty-six for women, all without the kind of happiness that the *Hello Kitty* brand brings.

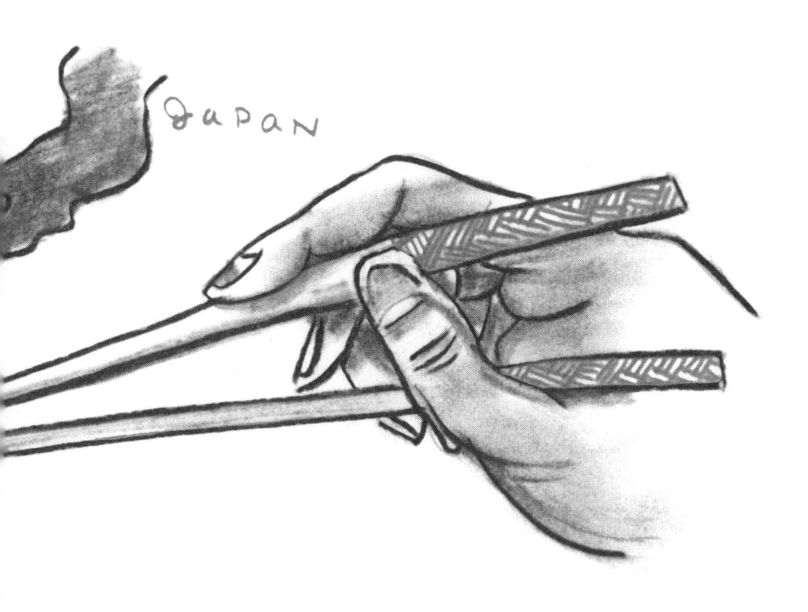

Age-Old

By 2030, almost three fourths of all eight-year-olds in America will have one of these. What would that be?

Despite suffering from three times the dementia, four times the breast and prostate cancer, and five times the heart disease of Okinawans, Americans are still surviving—more than ever! After leaving Okinawa and World War II, we traded B-29s for a baby boom; today, according to research at Berkeley, we're in the midst of a great-grandparent boom. With the American life expectancy nearing seventy-eight and a 35 percent increase in centenarians in 1990s, it is estimated that by 2030, 70 percent of eight-year-old kids will have a living great-grandparent. (These statistics will make a lot more sense if you remember that a centenarian is not a half-human/half-horse crossed with an animal doctor but someone who lives to be 100.)

A Snail's Space

Why are the **Chinese** so interested in watching snails from space?

The Feeling Sluggish Answer

At times, it seems as though everything is bigger in China: the 4,500-mile-long Great Wall, the 1,614-foot-high Shanghai World Financial Center, the 7-foot, 6-inch Yao Ming. In fact, two thirds of these are visible from space—well, only one third when the Houston Rockets star is lying down. Also visible from space, thanks to new European satellites, are snails, which, like the average Chinese person, are quite small.

From the Three Gorges Dam in Shanghai (no, not the "Three Gorgeous Dames in Shanghai," the massage therapists), satellites are monitoring water levels and conditions of the country's largest body of freshwater, Poyang Lake, to determine whether it is inhabitable for parasite-carrying snails. The snails (nowhere near as tasty as escargot) host schistosomes (nowhere near as nice as that sounds, and it doesn't sound nice). These cause schistosomiasis, a chronic disease that can lead to bladder cancer, kidney and liver damage, and blood infections, along with the impairment of juvenile growth and development.

The second most prevalent tropical disease (malaria comes in first), schistosomiasis is contracted from contaminated water and affects over 200 million people worldwide. Along with calculating water temperature and other factors affecting the billions of chickpea-size snails in Poyang Lake, European Space Agency satellites measure the lake's water level by bouncing radio signals off it and tracking how long they take to return.

Clearly, satellites have a lot of time on their hands.

Pardon Our French

Compared with the British, the French do this 142 more times in an hour. **Do what?**

The Touch-and-Go Answer

We keep in it, we're out of it, we lose it, but the French, as a curious bit of research reveals, do an awful lot of it: They touch one another 142 times in an hour, whereas in the same hour a Londoner is unlikely to manage a single instance of hand-to-hand contact—or hand-to-arm, hand-to-face, or hand-to-any-part-of-another-person. Touching takes many forms, and someone obviously wrote a big grant (perhaps "Disambiguation of Types of Exhibited Tactility") so that we can better understand the five types of touch:

- **Functional:** to accomplish a task in an impersonal and often professional setting. Examples include an uncomfortable patdown by a nightclub bouncer or a bartender's pat of your hand, which increases the chances you'll tip her (according to some other studies we don't have time to touch on).
- **Social:** impersonal yet polite gestures that establish a welcome connection or a mutual sense of confraternity, such as slipping that bouncer a Ben Franklin to get into the VIP lounge (now that really hurts).
- **Friendly:** Indicates a personable, platonic relationship with another and includes high fives, noogies, wet willies, arm burns, wedgies, atomic wedgies, and other equally affectionate interactions between kids.
- **Love:** intimate touch indicating emotional attachment, including strong sustained embraces with another person or, depending on your politics, hugging a giant redwood, the gates of the White House, or the goalposts of the winning team.
- **Sexual:** for stimulation and to show physical attraction.

Despite all the butt slapping after the big game and the hand shaking at press conferences and even Obama-esque fist bumping, studies show that American adults engage in physical contact less frequently than most Europeans and Latin Americans—proving that, as we age, most of us lose touch.

A Mound of Trouble

Unless you're looking for a lifetime of trouble, you'd better not do this more than 1,000 times per season and no more than 3,000 times in a year. **Better not do what?**

Bone-Plate-Versus-Home-Plate Answer

Little League baseball games are rife with hazards: unstable bleachers, unstable beverage caddies, unstable parents. The events *on* the field are dangerous, too, with all those wild pitches, far-flung bats, and eyes-on-the-ball center field collisions. With year-round games and the pressure to win soaring higher than a fly ball off an aluminum bat, your aspiring pitcher is increasingly likely to need reconstructive surgery, specifically an elbow operation on his ulnar collateral ligament, known as Tommy John surgery.

Wear-and-tear trauma used to be reserved for major leaguers. Before he underwent the operation that now bears his name, John had pitched for various teams for more than a decade. Although he spent the 1975 season recovering, he then took the mound for fourteen more seasons.

Nowadays, record numbers of young Cy Youngs, some only ten years old, go under the knife.

Why is this? Kids are now more likely to specialize in a single sport (another reason to take up ballet in the offseason), and the trend toward longer seasons leads to young pitchers overusing their arms: more games, more pitches, more stitches. In 2007, Little League baseball limited all ten-and-under players to seventy-five pitches per game, eleven- and twelve-year-olds to eighty-five, and all fledgling hurlers to fewer than 1,000 pitches per season and 3,000 per year. (Oddly, they did not address the need to regulate Pixie Stix, Mountain Dew, and other such performance enhancers.)

Young players' bones are especially susceptible to injury during puberty, when repetitive, strenuous movements can harm nearly closed growth plates. Score another one for ballet: A plié isn't exactly strenuous.

Unfortunately, most youngsters would prefer a grand slam to a grand jeté. (Do dancers get trading cards? You bet they don't.) In one study, pitchers who routinely threw despite arm fatigue were thirty-six times more likely to need the surgery. When they find out that the recovery time is six months to a year, outraged parents know there's only one thing that can still advance their child's quest to win the Little League World Series: forge a birth certificate and let them remain eleven years old for the next few years.

Frosty Reception

Let's say you're feeling out of place: Strangers at a party eye you from a distance, you disappoint your bowling team with the lousiest set of the season, nobody's chatting you up at Thursday night speed dating. In these left-out and left-alone scenarios, which are you more likely to crave:

a) A bowl of tomato soup
b) A cappuccino
c) A diet soda
d) A Granny Smith apple
e) A handful of crackers

The Out-in-the-Cold Answer

Sure, an ice-cold beer could enhance your schmoozing, but how do we know you're of drinking age? So let's go with recent research: You're more likely to long for a cappuccino or soup, or any hot beverage or food, because they tend to alleviate your social discomfort.

We often say "icy stare," "chilly reception," or "cold shoulder" to describe someone whose social demeanor makes us shudder. Well, it turns out that these temperature-related idioms are more than just metaphors. Those who feel socially ostracized actually feel physically colder—and not just because the cool kids wear super-warm, super-trendy earmuffs.

In a new study, one group of students was told to remember a time when they felt socially included, and another was told to recall a time when they felt excluded. When asked to estimate the room temperature, the students' responses ranged from a wintry 54°F to an equatorial 104°F. (Fifty degrees is a pretty wide margin of error, even for humanities majors.) Overall, those with memories of exclusion estimated that the temperature was 5 degrees lower than the group remembering warm and fuzzy feelings.

Scientists attribute these reactions to the insula region of the brain, which tracks both body temperature and psychological mood. Guess that means cool people come with better "insula-lation."

In another experiment, participants played a game of catch on a computer, ostensibly with other students, but it was the computer that either threw them balls or ignored them. Afterwards, when offered a variety of foods and drinks, the nonreceivers clearly preferred hot soup and coffee. The receivers showed no such preference, and it's not because they worked up an appetite running across the field.

The lesson? Here's one more reason to pass on the rocky road ice cream when you get dumped. Mom was right about chicken soup warming the soul. Thanks, Mom.

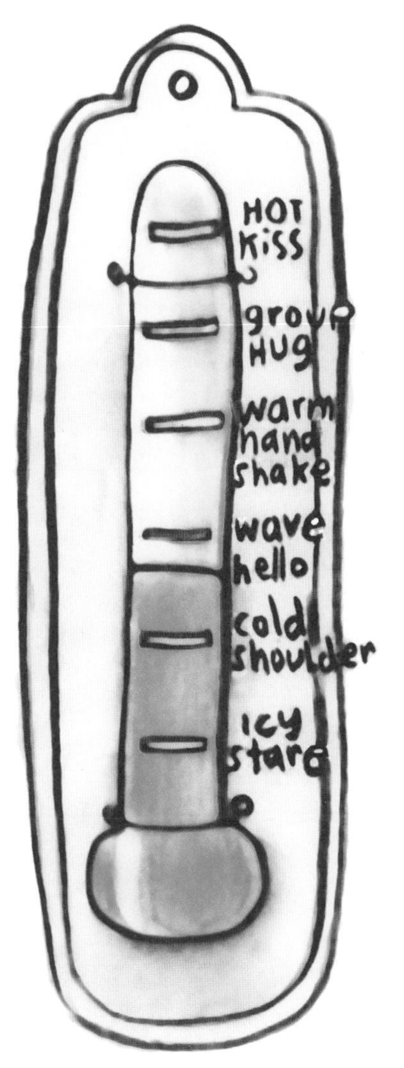
HOT
Kiss
group
HUG
warm
hand
shake
wave
hello
cold
shoulder
icy
stare

Tips for Your Next Cool Party, Hot Party, or Nice Warm Party

Help your guests break the ice:

- **Serve hot food and beverages so everyone feels comfortable.** Research shows that even holding a hot drink in your hands can make for warmer feelings about the people around you. (Note: Although Hot Pockets do count as hot, they do not count as food.)
- **Alcohol may free up all those inhibitions and loosen the tongue, but it actually decreases core body temperature.** And, as we have seen, cold temperatures inside and out can make people feel less cozy, more isolated. Also, keep the party optionally virgin with alcohol-free drinks on hand.
- Especially in the winter, when cold temperatures contribute to seasonal affective disorder, light a fire or jack up the heat to make your guests feel less SAD.
- **But don't turn up the thermostat too high, which can make your guests crabby or lethargic.** Worse, you could trigger their internal temperature regulator: When your brain gets too hot and your body needs to cool it down quickly, you yawn. Not exactly a come-on signal.

Going Up?

Gravity is a formidable force, **of course,** but when it comes to pure power, what pushes just as well when you stand on your feet as when you stand on your head?

The Answer to Get You Back on Your Feet in No Time

Your heart keeps pumping whether you're upside down, right-side up, or head over heels; however, gravity helps your heart keep things circulating, and if you're upside down your blood just doesn't get around much anymore.

But one thing you can do well upside down is swallow. That's right: Eat! In the esophagus, contractions called peristalsis push breakfast, lunch, dinner (and, for college students, the inevitable fourth meal) down the 10-inch digestive chute like a parent encouraging a timid child down a playground slide. (Your doughnut's next ride: the see-saw! The duodenum is fun!) Those contractions are so forceful they overpower gravity and allow you to swallow up, digesting even if you've inverted yourself.

Sure, you can hang sloth-style from a tree, but it's not recommended that you sleep upside down, for you are not a bat or Batman—or the original Robin, who came from a family of circus acrobats. Remaining inverted for too long can cause a harmful pooling of blood in your lungs and head. If you have high blood pressure, that could cause a stroke. As for "inversion therapy," it's been proven ineffective for long-term relief of back pain. A better idea is a simple yoga headstand, *sirshasana*: It strengthens your neck and upper back, increases coordination and body alignment, and has a calming effect on your thinking (but don't do it if you're pregnant or your doctor diagnoses high blood pressure or certain eye diseases).

Even David Blaine, the public relations magician, couldn't complete his three-day upside-down hang-a-thon without receiving medical checks in the upright position once an hour. Plus, not peeing for three days? Now *that* would have been magic.

Nuts and Volts

Here are your choices: Stand on a ray, wear one on your forehead, or hold an electric catfish. Obviously, something is wrong with you. What?

The Elec-trick Answer

You're an anachronism. You're in ancient Rome. (And you're not there just to ride the waves on your new ray board, to flaunt the most shockingly odd headgear, or to buy your daughter the electric catfish she'd been begging for since her last birthday.)

When in ancient Rome, do as the Romans do; sometimes it's something shocking.

Today we have precisely calibrated medical machinery (and sophisticated medications to make it possible for some people to avoid machinery altogether), but in the old days a doctor sometimes prescribed a jolt from an electric eel, catfish, or ray. This early form of electroconvulsive therapy (ECT), also known as electroshock treatment, was used by the Romans to treat the joint pain and inflammation caused by gout. The shock docs also used rays to ease headaches (they'd apparently never heard of a nap or ibuprofen) and to treat seizure disorders such as epilepsy.

The modern form of medical electronics began in the 1930s with German psychiatrist Hans Berger, who was described as "a modest and dignified person, full of good humour, and as unperturbed by lack of recognition as he was later by the fame it eventually brought upon him." This humble health care professional would have been ideal if not for a few minor limitations: He didn't know anything about mechanics or electricity. But his patients and colleagues trusted him enough to let his shortcomings (electric) slide, and his practices continued and evolved into present-day treatments.

ECT is used when other treatments fail to alleviate severe depression. To begin with, the patient receives general anesthesia and an immobilizing muscle relaxant. Then doctors administer up to 450 volts to induce a 30- to 60-second seizure; in certain cases, patients receive treatments three times a week for about a month—sort of like your exercise routine after the New Year, only you get struck by lightning every time you step off the treadmill. Side effects of ECT can include permanent memory loss and cognitive impairment. But really, who would want to remember being zapped by one-fourth the voltage of an electric chair?

Frown to the Last Detail

To improve your first-grader's ability to complete detail-oriented school assignments, you should offer the child:

a) A monetary bribe
b) A munchable treat, such as candy
c) The chance to watch an animated film in which a sorrowful fate awaits a cheerful little character

The Depressing Diagnosis

If you choose (a), your child will probably be expelled after his newfound cash flow leads him into the shadowy world of illegal juice box trafficking. Choice (b) can only result in smuggling Butterfingers into gym class, punishable by the child's version of the big house: the chair outside the principal's office. Moreover, neither of these bribes will help your child focus as well as (c). Studies show that children in a neutral or pensive mood are more attentive to detail than exuberant ones.

In one experiment, six- and seven-year-old kids watched either a happy, neutral, or sad scene from an animated film and were then told to identify figures embedded in a complex picture. Compared with the cheerful guinea pigs, children who reported feeling sad or neutral discovered an average of two or three more figures.

The reason, researchers say, is that happiness signals a sense of personal safety that encourages a relaxed, broad focus on one's immediate situation. Sadness, on the other hand, makes you aware that a situation is difficult and predisposes you to concentrate on details needed to solve a problem. The same applies to adults: Slightly gloomy grown-ups performed better in tests of social judgment ("Can I really eat ribs with my fingers? in public? without a bib?") and visual memory ("Our honeymoon? Of course, I remember! You wore, like, this really nice dress-thing, and then you took it off, and. . . .").

And although these temporarily sorrow-induced people test better, the studies don't necessarily reflect a person's overall state of mind. Happiness improves creative thinking; students perform better when they enjoy school. Also, happy kids learn better in general—even if they miss more details—so don't rule out candy and money just yet.

It Really Is Next to Godliness

You're serving jury duty. It's time for the verdict. You'll feel more inclined to hand down a decision of not-guilty if you've just done what?

The Squeaky-Clean Answer

In the famous scene from *Macbeth,* Lady Macbeth, overcome with guilt after convincing her husband to commit murder, tries to scrub imaginary blood from her hands while crying, "Out, damned spot!" Until recently, Shakespeare scholars have viewed this as a metaphor for ridding herself of her guilt, but now we know that our brains react the same way to the unclean feeling of dirty hands and the unclean feeling of a dirty deed. (Too bad for Lady Macbeth that purse-sized wipes weren't available for nine more centuries.) New evidence suggests that when we feel physically clean we're less judgmental. The "Macbeth effect" conflates cleanliness and moral purity.

In a recent study, subjects watched rather disgusting clips of the drug film *Trainspotting*. Those who were instructed to wash their hands afterwards were less harsh in judging the transgressions of the film's characters. The same goes for real life: Voters who scrub-a-dub-dub before going to the polls are more likely to forgive a potential candidate's political misdemeanor. Watch for legislation introducing soap and water outside voting booths.

In another trial, participants had to unscramble 40 four-word sentences and then rate moral dilemmas such as keeping money found inside a wallet, lying on a résumé, and using a kitten for sex. The participants whose sentences had included words such as "pure," "washed," "immaculate," and "pristine" were more charitable. On a scale of zero ("perfectly acceptable") to nine ("extremely wrong"), those who unscrambled the clean words rated an average of 0.8 lower than the other group. In the most startling disparity, the "clean" group rated the "sex kitten" at 6.7, whereas the others scored that an 8.25.

What do you say we all hit the showers?

Your Money and Your Life

Currently, Americans spend about 847 million hours each year—**an average of 70 minutes**, every single week, for each of us adults—doing what?

The Stitch-in-Time Answer

Americans fracture about 5.6 million bones per year. One quarter of us eat fast food every day. An undisclosed number of teens fake stomach cramps, headaches, and other boo-boos to avoid gym class or algebra.

All told, we spend 847 million hours annually receiving medical care. (These hours factor in drive time, waiting rooms, examinations, treatments, pharmacy visits, bill paying, and schlepping others to their appointments. Time spent grousing about all this is not included.)

And we can't blame hypochondriacs for throwing off the curve. People aged sixty and over spent twice as much time receiving health care as those between fifteen and sixty, and despite spending more time on exercise, personal hygiene, and feeling everyone else's foreheads, women spent 70 percent more time than men obtaining medical care.

While we as a nation pay for more insurance premiums, co-payments, deductibles, and noncovered services, Americans still continue to seek more health care. Along with the increase in demand for those outdated copies of *National Geographic* and *Meet Your Spleen,* there's stiffer competition for one of those uncomfortable waiting room chairs. (If you thought a waiting room would be more accurately described as primary-colored foam furniture, dog-eared issues of *Highlights for Children,* and learning toys scattered in front of a continually running DVD of *The Lion King,* it's time to graduate to a big kids' doctor.)

More and more, doctors' offices resemble overbooked commuter flights; in fact, a fourth of all patients wait more than half an hour to see a doctor. Nearly 20 percent wait more than a week for an appointment.

Doctors are complaining also. Patients procrastinate until symptoms are serious, they neglect to mention potential problems (such as chronic severe headaches), and they expect the doctor to diagnose them quickly and accurately when nonspecific complaints ("I feel achy") can mean anything from "I have a psychiatric disorder" to "I danced all night in brand-new three-inch heels" (see "psychiatric disorder"). And only a sixth of general practitioners' office visits are with patients whose bodies need medical attention; the rest suffer from psychological or stress-related conditions (i.e., spending seventy minutes of your week waiting for the doctor to see you).

Saltwater Crocodiles, Seals, and You

In a given year, the average woman does this more than three times as often as a man, but the average girl does it just as much as a boy. **What is it?**

Blood, Sweat, and Answers

Another thing women do 3.8 times as often as men: treat themselves to parfaits, subscriptions to bridal magazines, and $100 facials. Whether a woman realizes how many calories she consumed in just one parfait, the fact that her wedding dress budget might not even cover the price of a Vera Wang veil, or that an expensive facial still doesn't bring back youth, they all prompt the answer to our question: crying. In a given year, women cry sixty-four times, whereas men cry seventeen times. (Men also punch their fists through more walls than women and tend to sweat more. Their perspiration contains toxins that are otherwise released in tears.)

Although Charles Darwin described tears as "a special expression of man," scientists and psychologists have several theories on why tearing up is more common for women. Women tend to cry when sad (i.e., three months after going to homecoming, "I'm pregnant!") or happy (i.e., three months after coming home from the honeymoon, "I'm pregnant!"). Men tend to cry when expressing negative emotions (i.e., "How can you be pregnant?"). Another theory suggests that evolution provided different structures for men's and women's glands, the lacrimal (or crying) glands in addition to the mammary ones.

Everyone sheds the same three types of tears, and yet, like fingerprints, each person's tears are unique.

- **Basal tears** are the 5 to 10 daily ounces of lubrication we naturally produce to keep our eyeballs moist and comfortable, protected from foreign particles and other airborne irritants. You are free to blame your teary reaction at the end of boy-and-his-dog movies on your basal tears. Feel free, but no one will believe you.

- The tears that well up when the eyes are irritated by smoke or chopped onions are called **reflex tears** and are released when the brain stem sends soothing hormones to the cornea. Thanks to modern medicine, doctors don't test the reflexes of tears by striking the eye with a tiny mallet.

- Whereas reflex tears are made up of 98 percent water, the third type, **emotional tears,** have a distinct chemical makeup that includes leucine enkephalin, an endorphin. That's why tears make you feel better, but nothing is going to make you feel all better after that boy had to give up the only one in the whole world who truly understood him, his puppy.

After puberty, hormonal changes drastically affect crying for all of us. Women develop vast reserves of prolactin, a hormone that balances body fluids and, as the name suggests, regulates milk production. A build-up of prolactin can trigger weepiness.

As for the title's mention of saltwater crocodiles and seals: Those creatures also produce tears, although not the fake kind that accompany acceptance speeches on award shows. Indeed, if we were sea dwellers at some point in our evolution, our tear ducts would have functioned as a means of eliminating salt from the body. How tears became associated with squishiness and emotions is any body's guess.

Cry Me a River

Culture has always played a significant role in crying. In eighteenth-century Europe, cultivated men were known for their sensitivity and cried openly in public, especially at the opera. Now crying in public is the privilege of men strong enough to be taunted as "big fat babies."

We all cry at different times for different reasons, and some of us cry for no reason at all. Here's a roster of the great weepy wonders of the world.

Town Crier	Tearful Moment
Mourners in ancient Greece 	They bottled their tears to bury them with the departed as a demonstration of their great grief. (Seems strange until you realize that today we often bottle the departed themselves.)
Actors and models 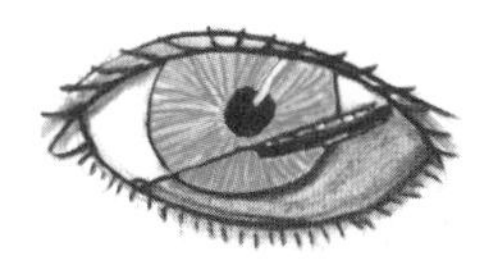	When a job calls for weeping, they resort to thinking of sad thoughts or smearing camphor or menthol around the eyes. Phony tears require the strategic application of glycerin or contact lens drops on the cheek. What also works: imagining a life without makeup, airbrushing, and diet pills.
Iron Eyes Cody 	Italian American actor who portrayed the crying Native American on the 1970s *Keep America Beautiful* TV ad. His tearful trick? Simple: Imagining the smoke-filled skies above New Jersey.

Town Crier	Tearful Moment
Anat 	According to the Ras Shamra texts, Anat wept upon hearing of her brother's death, the first record of tears in history. Apparently, even Adam remained dry-eyed when his rib was taken to create a mate for him.
Soldiers in World War I	The effects of tear gas contributed to the prohibitions of the Geneva Protocol.
Lou Gehrig 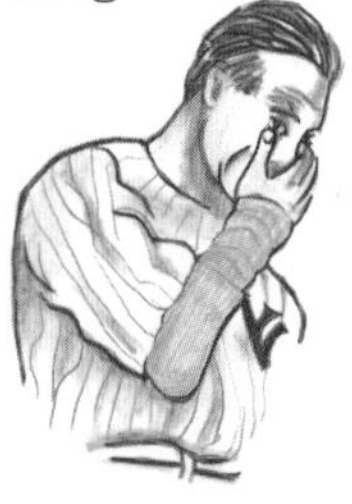	On July 4, 1939, Gehrig retired from the New York Yankees because of the disease that now bears his name, and yet he pronounced himself "the luckiest man on the face of the earth." Obviously Gehrig's misty speech predates the 1992 film *A League of Their Own*, in which Tom Hanks's character proclaims, "There's no crying in baseball!"
Bob Marley 	Word has it that his 1974 hit "No Woman, No Cry" was originally titled "There's No Crying in Reggae!"

Feeling Nosy

You've bought a unisex perfume gift set for your love. **It's Valentine's Day!** If you're a match destined to be "genetically successful," one or the other of you should smell what?

Sniffing Out the Answer

First, you should realize that unisex perfume is a pathetic gift. You chose *that* over the hot-air-balloon-sized teddy bear, the trip to the moon (yours for three easy payments of just $33 million), and the therapeutic canine sauna that popped-up in an ad on a Web page?

If both of you appreciate that an evening clinking champagne glasses in an actual hot-air balloon makes a much better present, then you and your partner are off to a good start. But seriously: A sure sign that you're a perfect biological match is if one of you can't stand the perfume's scent.

The reason, as scientists explain in words we've tried to understand more than once, has to do with our major histocompatibility complex (MHC), the region of the human genome responsible for partially coding our immune system. Evolution has taught us to mate with those who have an MHC with a different chemical makeup from our own. Why? Because mixing two MHCs boosts the offspring's immune system. This is yet another compelling reason to consider your foxy first cousin *totally* off limits.

You're probably wondering where smell comes in. Humans are able to smell the MHC and, unlike the sound of our own recorded voice, we actually like our own chemicals. Therefore, we buy perfume or cologne that advertises our own scent, and we're less inclined to like the odor of someone with a different MHC. Thus, in a genetically magnetic couple—opposite MHCs attract—a unisex perfume should smell like precious sandalwood or whispers of lily to one and like old sandals or Grandma Lily's bed sheets to the other.

So what's the best way to get your Valentine's pheromones moaning? Spritz yourself with a cologne that you—not your significant other—find attractive. In fact, cologne can boost a man's self-confidence so much that women who merely observe him (no smelling involved) will find him significantly more attractive.

And if your new scent doesn't work its charms, remember, your Grandma Lily will still think you're the nicest of all her grandchildren.

Spicy Bonus with Hints of Woodiness

Perfume critics use scores of evocative words—*peppery, green, ripe, earthy, juicy*—to describe odor, similar to the way their often-mocked wine critic colleagues go on and on about grape juice. How can you describe the top notes of a fragrance's first whiff to the bottom notes of its last waft and still have someone's attention?

One way to talk about aroma is to use the fragrance wheel. It's similar to a color wheel in art. At present, there are five "notes" on it: floral, oriental, fresh, woody, and fruity. (Time for a new Spice Girls group?)

Also notable is where fragrances come from. These days most are compounded from synthetic odorants and plant sources. But we have not always been so green. Here's where some common perfume ingredients originate:

Castoreum Produced in the castor sacs (don't ask) of beavers, castoreum is dried and used to create a fragrance's leathery notes.

Amber This is actually ambergris, not the yellow stuff that traps insects and hardens into jewelry. Prized for its earthy aroma, ambergris is secreted in the intestines of sperm whales and can be found floating in the sea. (Yes, in the bad old whaling days, we went about getting it the hard way.)

Musk Originating in glands located in the nether regions of the male Asian musk deer, this is now produced synthetically and known as white musk.

Civet From the animals that brought you "dung coffee" (don't ask) comes a greasy secretion derived from a region that a Depression-era flapper would call "the last thing on a cat."

Honeycomb Honeycomb contains two substances—honey and wax—that have been used in perfumes worldwide since the dawn of vanity. Besides their delightful smell, these substances can be pinched with minimal inconvenience to the animals that make it.

Face It

What makes up only 21 percent of your face but takes up 45 percent of our attention?

The Eye of the Beheld

At some point, you might be asked, "In rating someone's attractiveness, what's more important: the face or the body?" to which you should immediately answer, "Personality." You might be lying, but some things hold true. A body can be masked in heavy clothing and a personality in heavy alcohol consumption, but a face always shows face, and our eyes are drawn to the eyes.

When it comes to facial features men find most attractive, many would still say, "The breasts." Women, being much more evolved, insist on intellectual stimulation—you know, someone intellectually tall, broad, and ripped. Despite personal tastes, when any of us look for clues to a person's age or health or even their personality, we look at the eyes. In a recent study, people were shown photographs of older people and asked to rate their ages as researchers tracked where and for how long the participants' pupils focused. For nearly half of the time, participants gazed at the area around the eyes. The second most examined feature after the window to the soul was the window to the sinuses, or the nose (just under 20 percent), followed by the window-that's-really-a-wall-to-the-skull, the spot above the eyebrows (just over 10 percent).

Although the eyes take up a mere 21 percent of a face, they drew the most attention (44.7 percent) when participants rated the tiredness of the photographed subjects. Based on the study, researchers say those wanting youthful and vigorous looks (their own) might consider cosmetic eye surgery. (It's hardly surprising that the researchers recommended surgery because the goal of the study was to determine which surgery was most advisable to people who want to look younger.) A more frugal option is to wear sunglasses day and night and to accentuate your remarkably young-looking nostrils.

Don't bother with plastic surgery if you're trying to impress someone with prosopagnosia; they won't recognize you either way. One in fifty people suffers from a mild form of "face blindness," which prevents them from recognizing other people's faces, and sometimes even their own. There is no treatment for this socially awkward condition; some are born with face blindness, some develop it after a head injury, and some simply can't recognize their own mother's face after the Botox, chin tuck, and eye lift.

Shoots and Ladders

If you're one of those mothers who wants her son to be a future NBA all-star—**and what mother isn't?**—what's one step you can take even before your baby boy takes his first step onto the court? (Works for daughters hoping for the WNBA, too.)

The Lying-Out-by-the-Gene-Pool Answer

Try playing him clips of last season's all-star game through the ultrasound. Or try painting a giant number 23 on your belly and naming him LeBron. While you're deciding which of these is most unreasonable, begin saving for his college fund. Then, you can begin praying he'll reach the average NBA height of over 6 feet 7 inches tall. (Because what's more comforting than an endorsement deal and a mansion for Mommy?)

Mothers, it's time to hit the beaches or take vitamin D. That's right: Pregnant women who are exposed to sunlight see their children grow taller. (Plus, a tan belly can be decorated as a jack-o'-lantern for a really eye-catching Halloween costume. We're full of suggestions! Just ask.)

An eighteen-year study of more than 7,000 children showed that vitamin D, which the body cooks up from the sun's rays, helps babies grow stronger, wider bones.

Like any good thing, too much of it is no good: Excessive exposure to UV rays can be harmful to pregnant women and the rest of us. Most older, bigger babies (the kind known as "adults") receive their minimum requirement of vitamin D by eating a healthy diet and getting five to fifteen minutes of exposure to the sun each day to help decrease the risk of a fracture or osteoporosis. No need for tanning beds, skimpy clothing, or vacations on undiscovered little tropical islands (most of which are owned by earlier generations of famous NBA babies).

So, moms-to-be, get your daily allotment of vitamin D by basking poolside as you prepare for many years of sleepless nights thanks to your darling infant, lanky adolescent, crazy knucklehead, and finally basketball star. You're going to earn that mansion.

BUST
NBA OR BUST

for D Students

Vitamin D helps your body prevent certain cancers, staves off the symptoms of Alzheimer's, and absorbs calcium, which increases bone mass and fortifies tooth enamel. (If conversation languishes, go ahead and gaze longingly into your date's eyes at dinner and murmur, "You have really strong teeth." Said the right way, it can be really endearing.)

If you're a medical resident or a vampire and you don't get your minimum daily requirement via sunlight, you can fortify yourself with vitamin D by eating some foods, including foods that were fortified themselves (tarted up with added vitamin D): breakfast cereal (check the side panel), cheese, eggs, liver (you don't have to eat this in public), fatty fish (catfish, eel, mackerel, salmon, and tuna), and milk.

And there's always the popular standby, that old favorite of grandmothers everywhere: cod liver oil. Yes, it's still legal in fifty states. But do check the label; some brands actually *remove* the vitamin D.

A few more precautions: Because vitamin D is fat soluble, it won't just wash harmlessly out of your system. Excess amounts can be toxic, interact with medications, or create medical problems (such as nausea and confusion). Don't prescribe even "harmless" things such as vitamins for yourself; ask your doctor. And although sunlight may be your best bet for a free source of vitamin D, make sure you don't get so much protection from *some* cancers that you run the risk of *other* cancers (such as melanoma). In that case, the "D" in vitamin D would stand for "dumb."

Grease *Is* the Word!

Prevent weeds! Curb snail populations! Fertilize your garden! Prevent water from evaporating in your kiddie pool! Sop up that greasy spot in the garage under your car! **Do all this—and more—with what?** It's something almost everyone gives up all the time.

The "Salon, It's Been Good to Know Ya!" Answer

You—yeah, you, champ of the shampoo, commissioner of the conditioner, watching your stray hairs form a clump in the drain. (And what are you going to do with it—make a toupee for your Mr. Potato Head?) Wouldn't you rather contribute your lost hair to an organization that could use it to blot up some of the 706 million gallons of oil that enter the planet's oceans each year?*

An Alabama hair stylist started a program to collect hair clippings from salons and barber shops, and now more than 370,000 of them participate in the United States alone. The nonprofit, A Matter of Trust, collects about a pound per day from each member business to produce half-inch-thick, two- by three-foot mats of human hair. Though the perfect size for a welcome mat, this hair mass can hold 100 times its weight and slurp up oily blobs from oil spills, leaky cars, and salads you forget to order with dressing on the side. But wait, there's more! It can be reused more than fifteen times!

Human hair is adsorbent; this is similar to absorbent, but the first *b* is turned around, and the adsorbent hair gathers liquid around it, just as it gathers the oil our bodies make. (Which is why your hair feels greasy after you skip a couple of showers.) In fact, you're having your own slow-motion oil spill all the time, but your hair helps to keep it in check. Our hair works the same way with petroleum from a ship, and that's why, when it comes to protecting the environment, the mat made of salon sweepings acts as a hairy godmother.

*Where do these millions of gallons come from? Oil changes (363 million), ship maintenance and container washing (137 million), air pollution (92 million), large spills (37 million), and offshore drilling (15 million). Additionally, oil naturally seeps up from the sea bottom, contributing another 62 million gallons.

Go green as you go for that crew cut! Ask your favorite cutter to donate to A Matter of Trust at **www.matteroftrust.org.**

Locks of Luck

What do the following body parts have in common?

The tops of your feet
Your lower lip
The bottoms of your feet
The palms of your hands

And what do these body parts have in common?

Your face
Your chest
Your neck

Turn the page for the **answers**.

Answers

Your hair, the little follicle it calls home, and its very own oil (or sebaceous) gland, make up one of your many, many pilosebaceous units. These units are found everywhere on your body except the body parts listed in the first question, and they're most numerous on parts mentioned in the second.

Your body's own oil, called sebum, comes in two grades: virgin and, well, not a virgin. (No, even Rachael Ray doesn't have her own brand of EVBO—that's extra virgin body oil!) Its function is to moisturize the skin and hair. Too little sebum and you could use some jojoba oil. Too much sebum and you're probably going through puberty. And if your sebum is just right, you're probably Goldilocks.

Pump It Up, Way Up!

You have a puffy face, skinny legs, and brittle bones. **You're probably a who who's just returned from where?**

The Spaced-Out Answer

You might be a graceless model who's recovering from a tumble off the runway, a grandmotherly beekeeper who's just back from a run-in with the hive, or even a middle-school kid coming home from, well, anywhere.

Or you could just be an astronaut who's just returned from space.

Despite the fact that all astronauts pass a battery of rigorous tests to qualify for a space mission, they return to Earth in the worst condition of their lives—and not just the lactose-intolerant ones forced to eat freeze-dried ice cream for six months. Old or not at the outset, astronauts return from long voyages with muscle atrophy and other problems associated with advanced osteoporosis, a disease that plagues older women in particular.

Just four to six months in orbit can decrease a person's bone strength 14 to 30 percent, with an accompanying significant loss of bone density, especially in the hips. A zero-gravity environment significantly lowers the everyday stress on the bones. So, with no gravity up in the heavens to pull on the bones, the body is no longer prompted to maintain or refurbish them, so they degenerate.

Outer space's microgravity also fails to push fluids down in the body. This results in a swollen face, clogged sinuses, and a stuffy nose (yes, snot, like everything else up there in outer space, runs up). Even with the heart pumping hard, astronauts may also lose up to a liter of fluid in their lower extremities, causing those back on Earth to comment on their "bird legs."

But do astronauts get angry at such personal (and immature) jeers?

Right.

"I just came back from the moon," they have been known to reply. "How about you? What have *you* been doing lately?"

Drink to Your Health

To help your body fight diabetes, cancer, and cataracts, **how should you order your martini?**

The All-Shook-Up Answer

Wait, wait, wait. We're assuming that gin and vermouth help prevent these ailments in the first place, right? Well, according to a recent study, if you do *assume* that, it won't make an *ass* out of *u* and *me*—it could make you healthy, in fact.

Scientists have long known the health effects of red wine (good when taken internally in moderation but, when spilled on a light-colored rug, known to raise blood pressure and result in conniptions). Research also shows that martinis, compared with most alcoholic beverages, are high in antioxidants that help protect against the aforementioned conditions, especially if the bartender obliges you with two or three more boosts of antioxidants, the olives.

Your martini must be mixed just so. Tweak your British accent, shag your voluptuous accomplice, and order that martini shaken, not stirred—and, for pity's sake, not with some mortifying moniker such as "Candy Apple-tini" or "Lemon-Lick-a-tini." (All right, we admit that last part was more for our mental health than your overall health.) And although scientists aren't sure why martinis are so good for you or why stirred will leave you less cured, everyone else is too busy imbibing to be concerned either way.

Where were we? Ah, yes: health. During a sixteen-year study of healthy men, those who had two drinks a day were the least likely to have a heart attack, and those who drank none at all were the most likely.

And a seeming paradox arises in moderate drinkers: They are better off than heavy drinkers *and* nondrinkers. That's right: Women who have one drink per day and men who have one to two are actually healthier than people who never drink (and a lot healthier than people who drink too much or who don't drink all week and then "catch up" on the weekends).

And then there are the French. Despite lending their name to French bread (merely a means of conveying cheese or butter to the mouth), French fries (oil conveniently conveyed in double-fried potato strips), and French toast (egg-soaked, butter-crisped bread), they can proudly flaunt their "French paradox": Despite a diet of rich foods and a split of wine at dinner, they still have a low likelihood of heart disease.

Still, you're unlikely to be leaving your doctor's office with a nearly indecipherable prescription: martini dry with olives, *agita, ad libitum, cum cibo,* no refill.*

*The three Latin terms of your prescription mean "shaken," "use as much as one desires," and "with food."

Tater Troubles

You know you should be eating more fruits, veggies, whole grains, blah, blah, blah. A new study from Poland offers yet another reason why a diet of potato chips and French fries will do you in. What's wrong with your two tater treats?

a) The sodium content (salt!)
b) Greasy handprints left on your khakis (fat!)
c) Something else (what?)

Double Cheeseburger, Side of Carcinogens, and a Coke

Turns out, the worst part about that starchy stuff is neither the sodium content nor the greasy handprint left on your khakis. In 2002, the Swedish National Food Authority discovered acrylamide, a chemical associated with heart disease and cancer. It's found in about 40 percent of the calories we consume, in foods such as potato products, coffee, cereal, and chocolate—basically, all the good stuff.

Starchy foods contain significantly more acrylamide than their fried nonpotato counterparts. The onion rings at Fuddruckers—we know, that only *sounds* like a curse word—contain just 13 parts per billion (ppb), but their French fries had 452 ppb. Even more acrylamide-abundant were Popeye's fries at 1,030 ppb. And although all fries contain some of the chemical, its prevalence is affected by variations in the natural components of raw materials and cooking conditions. These can cause varying acrylamide levels even at different locations of the same chain restaurant. For example, McDonald's fries ranged from 115 to 497 ppb at different locations, an inconsistency that's consistent with how often they get your order right.

Another varying factor in the argument over acrylamide is how detrimental it is to your health, if at all. U.S. Department of Agriculture studies showed that high acrylamide intake increases one's risk of heart disease. An eleven-year Dutch study of more than 60,000 women showed a correlation between a high-acrylamide diet and postmenopausal endometrial and ovarian cancer. But, talk about a second (and third and fourth) opinion: a study in the *British Journal of Cancer* showed that acrylamide may actually reduce one's chances of large bowel cancer, while the *European Journal for Cancer* found no link between acrylamide and men's risk for colorectal cancer, and the *International Journal of Cancer* found that fried and baked potatoes have little association with cancer risks.

Hmmm. Our conclusion: Folks at the world's cancer journals aren't going to give up their tater treats. Meanwhile, think golden yellow rather than golden brown when roasting, toasting, grilling, or otherwise coloring your carbs.

Killing Me Softly?

Of all the dangerous animals on Earth—**the ones with mauling claws, venom-injecting fangs, flesh-ripping jaws, killer instincts, or remorseless aggression**—which one presents the most trouble for people in the United States?

The "Sting, Where Is Thy Death?" Answer

You're thinking of creatures with rapacious appetites, erratic moods, and unpredictable behavior: sharks, bears, and suburban teens. Or maybe you're guessing crocodiles, pit bulls, or gambling addicts.

Turns out, the country's most deadly beast is half an inch long, grows hair on its eyes (no, not a micro-werewolf), and prefers flowers to human flesh.

Yes, it's the bee, sweet symbol of industry, creator of honey, inspirer of the yellow-and-black outfits worn by cutesy trick-or-treaters and Pittsburgh Steelers. (And the yellowjacket wasp, a close relative of the bee, kills and injures three times as many people as any North American snake.)

A bee, which can fly as fast as a mall security guard on a Segway (about 15 miles per hour), becomes aggressive if swatted. And if hit, it summons others by releasing an odor that spells *ATTACK* in bee scent. Other factors that attract these belligerent buzzers include body odor and dark colors, making Amish funerals common targets.

Most bee stings, though painful, aren't deadly. For 99 percent of us, a fatal dose would be about ten stings per pound of body weight—that's 3,000 stings for a football lineman or about 90 for a supermodel.

But 1 percent of us overreact at times (yes, just like a teen), producing unusually high levels of the antibody immunoglobulin E. If IgE overreacts to bee venom, it floods histamines into the body's cells and causes anaphylactic shock: shortness of breath, nausea, inflammation and swelling (not just at the sting sites but in the lungs and throat), blue skin, and ironically—at least, when it comes to bees—hives.

So if you're the sensitive type, carry an epinephrine syringe (an EpiPen) to neutralize IgE and prevent a trip to the hospital—especially since one recent study in North America showed that only a small percentage of allergic emergency room patients received the recommended treatment protocol for bee stings. Ouch!

At least now you know the latest buzz.

The Lethal

Bees are the deadliest animals in the United States (and they kill about 400 people each year worldwide), but then again, we're short on lions and tigers. Thinking of the whole world, rate these four animals on a scale of 1 (cute and cuddly, like your cockapoo) to 10 (not so cute and deadly, like Cerberus):

a) Snakes
b) Elephants
c) Sharks
d) Mosquitoes

Turn the page for the **answers**.

Answers

Snakes: Despite the python and boa constrictor's hugging, most snakes get near humans only long enough to strike and kill more than 100,000 of us every year. Just imagine the pre–Garden of Eden world, when they still had legs! Our rating: 5.

Elephants: Though they seem lovable in the wildlife park or in pictures, elephants kill about 600 people annually. Our rating: 3.

Sharks: Sharks pick off only 30 to 100 humans a year, but that's mostly because we bail at the first notes of the *Jaws* soundtrack (*duh-duh, duh-duh*). Our rating: 1.

Mosquitoes: The deadliest animals in the world (thanks to malaria and all the other diseases they carry), mosquitoes kill more than 2 million people every year. Our rating: 10.

Of course, the old cliché is also true: The most dangerous animal to humans are other humans. And an EpiPen won't work on us pests.

Ready O.R. Not

You've just undergone surgery, and although you aren't awake to see this **(a very good thing)**, before closing you up, your surgeon passes a wand over your body. What sort of hocus-pocus is that?

Abracadabra: The Answer!

Gather up all the Monopoly money after your "millionaire volcano" erupts. Throw out the broccoli trees when you're bored with your brontosaurus village. Sweep up the shattered glass after hosting the first-ever indoor soccer tournament in the living room. One of the first things you learn as a kid is to clean up after yourself, yet even grown-ups forget, even the surgeons we trust to clean up our bodies after they make a bloody mess.

Along with leaving in needles and knife blades, surgeons have stitched up patients and left behind a surgical sponge souvenir during intra-abdominal surgery once for every 1,000 to 1,500 operations. During the game of Operation, it's easy to see the charley horse in your patient's thigh, but because a sponge is less conspicuous than a pencil or an ice cream cone or a pony, the O.R. tool kit now boasts a new radiofrequency detection system calibrated to locate not only sponges but also gauze, towels, and other lost surgical items.

Here's how it works: Each surgical accoutrement is marked with a bar code, and when the wand passes over the patient's body, it acts as a scanner. Picture the patient as a grocery cart in the "$10,000 or less" line. When the wand identifies a leftover article, a visual and audio signal notifies the surgeon, just like the little light bulb and the buzzer in the Operation game. Ah, the rewards of early training!

In some instances (1,300 to 2,700 times per year according to one report), doctors have been known to perform a "wrong-site procedure." In lay terms, that's an operation gone right but on the patient's healthy part. For instance, a team of Rhode Island doctors once operated on the incorrect side of a patient's brain, and surgeons in Florida amputated the wrong leg during a surgery. That's why they ask those annoying questions on your gurney ride to the O.R.—yes, you're *sure* the ingrown nail is on your left big toe. Remember, it's not rude to insist that your surgeon draws out your incision with a marker or to be firm about choosing a doctor with more operating experience than playing games in the family room.

Cooling Your Jets

Oddly, this is one of the things your body does that is contagious. In other words, if you see someone else doing this, you're more likely to do it—**unconsciously, not deliberately.** What is it, and what good does it do?

The Chill-Out Answer

Watching managers at a baseball game, you'd think spitting is contagious. Watching beauty pageants, you'd think crying is contagious. Watching stand-up comedy, you'd think laughter is contagious. The fact that folks have been imbibing for ninety minutes doesn't hurt.

For the rest of us, it's a yawn that spreads among a crowd as powerfully as a wave sweeps a cheering stadium. It's our body's one infectious function.

Admittedly, yawning is no asset when you're running for office and there's a camera nearby, if you're at a conference table and the boss is speaking, or if you're in a warm car on a hot date. In addition to boredom and sleepiness, yawning can communicate that your brain is overheated. When you yawn, you take a long, deep breath through your mouth and nose, providing your sluggish brain with oxygen and cooling it down; it's a little less of a wintry blast than shoving a peppermint patty up your nose, but it's more effective and less sticky.

Although yawning seems to indicate the body is succumbing to sleep, it is actually the way your body wakes itself up. That's why you yawn when you're nervous: It helps to perk you up and put you on top of your game.

In a recent experiment, participants watched clips of people laughing, yawning, or simply sitting there. Those who were asked to hold a 115°F hot pack to their foreheads yawned 41 percent of the time when they saw someone on the video yawn; those with a 39°F ice pack yawned only 9 percent of the time. Their conclusion: Cooler brains don't need as much yawning. (There's no data on how many people in the hot group felt feverishly delusional and how many in the cold group experienced ice cream headaches.)

Although yawning is associated with warm brains, preverbal communication in early humans, and the entire opera canon, there is also new evidence that our hard-wiring for empathy can also inspire a yawn. Without putting you to sleep with the details, preliminary findings suggest that one's susceptibility to contagious yawning correlates with how much compassion you feel for others. Therefore, a cool brain is not only an alert brain; it may also mean you can sympathize with the plights of others, even baseball managers, beauty queens, and hot heads.

Just to See if You're Awake

Students, assembly-line workers, museum guards—many people struggle to stay awake. You can use all five senses—and even sleep itself—to keep you awake. Try these pep-up tricks the next time you're at work dreaming of a fluffy pillow.

Winston Churchill, Napoleon Bonaparte, Eleanor Roosevelt, Bill Clinton, Salvador Dalí, Thomas Edison, John F. Kennedy—all of them **napped.**

Moving around increases your circulation (especially if you're moving in circles), oxygenates your brain, and gets you sharp. Get away from your computer, your workbench, or your confessional, and walk briskly, swim a couple of laps, or dance a mazurka.

Keep cool: Splash your face with cold water, drink a chilled beverage, open a window, or turn on a fan to give yourself plenty of fresh, moving air.

Light up your life: bright bulbs (full spectrum), neon signs (Vegas, baby!), and daylight, the brightest of them all. (If you're thinking tanning booth, we have a problem.) You need the sun to set your circadian rhythms, which have nothing to do with the cicadas that only set your ears buzzing.

Chew! It's what cows do with their cud, and they manage to stay wakeful through what must be the most boring lives on the side where the grass *is* greener! Chew gum or chomp on a small healthy snack. Just avoid sugar; it picks you up only to let you down quickly, without so much as an apology.

When you feel yourself start to nod off, **wake up by smelling the roses**. Or a dead skunk. Or, if you're Victorian, smelling salts (essentially, ammonia). A more popular alternative these days is aromatherapy: ordinary oils such as peppermint, eucalyptus, rosemary, or pine.

Coffee. Mmm! A little coffee will liven you up with caffeine; a lot will dehydrate you and make you less susceptible to its chemical charms. Wake up by smelling the coffee. Or vice-versa.

Drink plenty of water, fruit juice, or other noncaffeinated beverage. A full bladder will keep you awake.

Knead the muscles of your shoulders, neck, or lower back. Massage behind your knees or ears. See, there *are* socially acceptable ways to arouse yourself in public.

Fancy Ball Dress

Your boyfriend sewed a tennis ball onto the back collar of his pajamas and is drizzling honey into his mouth before bed. **Assuming he's not a superstitious person simply training for Wimbledon,** what on Earth is he doing, and should you try to stop him?

The Lie-Down, Roll-Over Answer

Hey, whatever goes on between you two is your own business, but if the tennis ball and honey aren't consensual, he's probably trying to make sure you have a good night's sleep.

Tennis balls, honey, giving up the late-night ice cream snack—these are all ways to prevent snoring. (Try telling your beagle that's not a chew toy on your boyfriend's flannel pajamas; now she *really* won't like the man.)

So what causes snoring, anyway, and how come almost half of all men and a third of all women saw logs while they're snoozing?

Your mouth has a blubbery roof (the soft palate) behind the bony roof (the hard palate over your tongue), and when you breathe air, it can make your soft palate vibrate. Now if your throat, nose, or mouth narrows or if there's any blockage—tonsils, mucus, or swollen glands—the possibility of making sounds is even greater. The narrower the passage, the more vibration. The more vibration, the louder the snoring. The louder the snoring, the more likely you are to be sleeping in the dog house. (Literally, since the beagle will probably end up sleeping in the top dog's bed.)

Sleeping on your back can collapse your air passage—just press your tongue on the back of your throat and try to breathe. Feel the problem? A tennis ball on your back collar makes it uncomfortable to sleep if you roll onto your back.

So why should your boyfriend give up ice cream (besides the obvious six-pack advantage)? Dairy products can create airway-constricting mucus. You don't see singers having milkshakes before a concert, do you? No. For them, it's a strict diet of hot water, lemon, and yelling (ever so quietly) at the stage manager.

Other helpful hints for snorers include honey and a humidifier (they both moisten the throat, making it less likely to vibrate). You might also try a small-breed dog you've trained to wake your boyfriend, a medium-breed dog to roll the snorer onto his side, or a large-breed dog to drag him out to the front lawn when he starts snoring. And then the tennis ball can be the dog's once again.

Just for Staying Awake

Or you can buy your boyfriend a didgeridoo (again, what goes on between you two is not our business). A didgeridoo is a long pipe made from a branch (usually a eucalyptus limb) that termites have hollowed out. It makes a better instrument than most other hollow tubes such as toilet paper rolls, bongs, or hollowed-out toy light sabers. (The Aborigines tried all of these before settling on their hollowed branch.)

The indigenous people of Australia play these "wooden trumpets" by blowing through the pipe with pursed lips, and the sound reverberates through the hollow tube. The trick is holding a continuous note, which is accomplished by circular breathing: breathing in through the nose and pushing air out using the mouth and the cheeks at the same time. Skilled didgeridoo players are able to renew the air in the lungs (keeping up the all-important oxygen exchange) while continuing to create sound in the didgeridoo for more than forty minutes without a single gap in the music.

Practicing this kind of breathing strengthens the muscles in the air passages so that they don't fall limp during a nap and start fluttering noisily. Unfortunately, while you master the didgeridoo, everyone in the house might well wish that you'd just fall asleep and go back to snoring.

The Truth, the Whole Truth, and Nothing but the Truth

How can you tell for certain whether your spouse is lying?

a) His mouth is open, and words are coming out.
b) She's perspiring, she's breathing quickly, and her heart is pounding.
c) Your spouse is taking a polygraph test, and the machine is going wild.
d) None of the above.

The "Liar, Liar, Pants on Fire" Answer

If you said your spouse is lying because (a) he's simply telling you something, this won't come as news: You need a new spouse. A little fib is one thing, but habitual lying in adults indicates deeper problems.

If you said (b), maybe it's your spouse who needs a new spouse. These are the same symptoms experienced by someone in distress. Not only that, you won't observe these indicators in a calm, cool, and collected liar—say, a professional crook or your congressman. These are also symptoms of illness or a high emotional state, or of someone who's just completed a marathon.

If you said (c) you're still incorrect. A polygraph doesn't detect whether someone is lying; it measures their perspiration, respiration, and heart rate, so this is the same as (b).

Which brings us to (d), the correct answer: "None of the above."

"Symptoms" of a guilty conscience, it turns out, are signs of unease, discomfort, embarrassment, anger, unhappiness—in general, the same indicators we'd exhibit if actually accused of something unfairly. As for shifty eyes and hesitant speech, fidgeting, and other sure-fire clues to guilt? Using these clues to assess the guilt of a complete stranger, you'd be accurate only 45 to 60 percent of the time. (By the way, you'd fare no better with an incomplete stranger.)

Not a very comforting average when you consider that we're all amateur seekers of truth and challengers of liars, or when you acknowledge that polygraphs and "truth serums" are administered with consequences that are literally life or death.

Even today, when scientists have isolated the brain's center for white lies, fibs, and falsehoods, the process of creating untruths turns out to be a curiously complex affair that requires some genuine juggling within the brain's centers that perform the conflict resolution needed to mask the truth. Even with the most advanced technology under ideally controlled conditions, scientists still can't say with 100 percent accuracy whether you're a faker, hypocrite, or phony or simply one of those people who can fool some of the people some of the time.

And that's the truth. No lie.

Open Wide and Swallow

If you ate an item from this menu—**small magnets, a long sword, locks of hair**—which would cause you the least amount of tummy trouble?

The Sharpest Tool in the Head

A hare, if cooked properly (may we suggest braised, with a classic *sauce moutarde?*), shouldn't make for stomach problems. On the other hand, hair, uncooked (may we suggest you keep your drains cleared?) can cause some genuine difficulty—especially in the case of a 10-pound ball of hair that had to be surgically removed from an eighteen-year-old with hirsute hankerings. (To help you visualize this, the operation removed a furry mass that resembled a large beaver.)

Then there's the girl who swallowed thirty magnets from a toy she'd received for Christmas. After surviving surgery to repair the eight torn holes in her intestines—you know what happens when opposites attract—even if, in the future, the kid got braces and a stocking full of oral bands and palette expanders, it still wouldn't be the worst Christmas ever.

No, the gentlest thing to swallow is the sword. Given that a sword is a sharp metal object generally used as a means of piercing, slicing, or flaying flesh, one would think more sword swallowers would become statistics in job-related accidents. All right, one Canadian sword swallower did die after swallowing an umbrella, but curiously, a sword can be quite the genial gastrointestinal guest. The sword swallower guides a sword down the throat by hyperextending the neck, aligning it with the mouth, and straightening the esophagus. Next, thanks to an ability to ignore the gag reflex and to relax some ordinarily involuntary muscles (those that open and close the esophagus and push food down it), the swallower moves the sword about 16 inches down the esophagus to the cardia, which connects the esophagus to the stomach.

Considering their line of work, professional sword swallowers suffer few injuries. A British survey of 110 of them (in the United Kingdom alone there are 46 professional cutlass-consumers) recorded the most common complaint as "sword throat," with symptoms like "your throat is so painful, it feels like you're swallowing a sword." (Similarly, Julius Caesar suffered from "sword back.")

Sword swallowing is not without its occupational hazards, such as a touch of intestinal bleeding or a small puncture of the pharynx or esophagus. The good news is that there were no deaths, and nineteen swallowers said they suffered no more than the occasional throat or chest pain. The risk of swallowing a sword once is low, but the odds of injury increase over the course of a long career, so all you sword swallowers should be sure to quit while you've still got a head.

Cough. [Cue the sound of choking.]

Clatter. [Cue the sound of the sword dropping to the floor.]

Leave It to the Professionals

Our vet has all too frequently offered us a term for when the retriever consumes most of a deer carcass or some other inappropriate vittles: The ensuing problems come from "indiscressive consumption." Here are some appetites that should have shown a bit more discretion:

- A French man died of complications from consuming 350 coins and other **miscellaneous metals**. For a man, a 12-pound mass that dislodges the stomach is called dangerous. For a woman, it's called twins.
- A woman went to the hospital complaining of abdominal pain and constant vomiting, and it was discovered she had ingested not a platter of spicy enchiladas but a **thermometer**.
- After a cockroach jumped into an Israeli woman's mouth while she was housekeeping, the woman used a **fork** to dislodge the pest and ended up swallowing both. Emergency laparoscopic surgery followed. People, remember your health science from eighth grade: Never put anything in your ear but a washcloth or your elbow.
- A twenty-eight-year-old man suffering from a personality disorder consumed a **pad of stationery**, and then, three weeks later, underwent surgery to remove two pens, a pen refill, and a pencil. He has had significantly more trouble finding a way to swallow e-mail.
- An older woman accidentally ingested her lover's **dentures** while trying out what she described as a "special type of passionate kiss." In this case, simply getting to first base brought up the "spit or swallow" question. She managed to pass the teeth without surgery.

Something to Chew On

Humans have survived for millions of years in a dog-eat-dog world **(not to mention lion-eat-lion, bear-eat-bear, and shark-eat-shark)**. Compared with these carnivores' __________, our own is rather small and feeble. What is it, and how have we overcome this shortcoming?

The Bite-Me Answer

We've long resided at the top of the food chain. Though well aware of our status, we've continued to prove this to ourselves by hunting the gargantuan (say, woolly mammoths with long kabobs called spears), the miniscule (say, escargot with tiny kabobs called cocktail forks), and the truly odd or offal (say, almost anything described as a delicacy).

Strange then, isn't it, that our jawbones—not to mention our stomachs, if you've ever been to one of the monkey brain–eating nations—are weaker than those of the majority of mammalian carnivores? Today, the animal with the strongest bite relative to its size is the Tasmanian devil, which has a bite force of more than 5,000 pounds of pressure per square inch. Just how devilish is this? The typical dog bite applies 200 to 400 pounds of pressure; a modern-day lion, 600 pounds; and a crocodile, 2,500 pounds. Although humans register only about 120 pounds of snapping pressure, you will have more respect for this ability if you visit a college pizza joint after midnight to see the species in its natural habitat during a feeding frenzy.

Human jaw strength has decreased over the eons while chimpanzees, who have learned to use tools, developed complex social networks, and mastered the driving of a car at least as courteously as most Bostonians, still have jaw muscles that are more than three times the size of a human's. Why is this so? Well, 2.4 million years ago, when we were not yet even sparkles in our parents' eyes, a genetic mutation left our ancestors with jaws far weaker than those of their prehistoric counterparts. The silver lining (as a matter of fact, it's actually gray, as in *gray matter*) is that this weaker jaw increased the skull's flexibility, which, in turn, allowed the expansion of the human brain.

So chimps got the better bite and we got the better brain, except for those of us with temporary crowns who tear into a pizza at midnight, giving some dentist's kid even better brains as we put them through college.

Bite-Size

Round

This country has seen incredible progress for those who like teeth that are healthy, white, and their own. As you dread your next trip to the dentist, try to remember you have less to dread than people at any time in history:

1. Before the advent of modern dentistry, other professions plied the trade. Which of these were considered to be "dentists" at one time?

a) Monks
b) Barbers
c) Blacksmiths
d) Silversmiths

2. Which of these has been used successfully to treat dental problems?

a) Egg shells
b) Sand
c) Clove oil

Turn the page for the **answers**.

1. The answer is "all of the above," even though we didn't give you that option. In the Middle Ages, monks were surgeons and dentists until their ability to practice medicine was drastically curtailed in 1215 by the pope. Then the barbers took over, and even blacksmiths were known to pull teeth (as were jewelers and wig-makers). At least one silversmith practiced dentistry: Paul Revere placed an advertisement promoting himself as a dentist. He was also the first person recorded to have identified a dead person by their teeth (and the dental work he performed).

2. Along with bones and myrrh, egg shells were used in Roman toothpaste recipes. Sand is baaaaaaaad for teeth and wears them down quickly, and the odd bit of grit in their food is probably why the teeth of ancient Egyptian mummies look as if they're chipped by faulty pyramid stones. Toothache sufferers have long used clove oil to treat problems from simple toothaches to serious abscesses. Even today, clove oil can numb a tooth or sore gum until you can get to the dentist.

See You in the Ring

Look at your index finger. **Now compare it to your ring finger.** What might you conclude if your ring finger is longer than your index finger?

Finally Fingered It Out?

Ever notice that men and women tend to examine their fingers in different ways? Women tend to keep their palms down and fingers outstretched, fretting over advancing cuticles and whether to choose "I'm Fondue of You" (milk chocolate) or "Bastille My Heart" (burgundy) nail lacquer. Men, on the other hand, keep their palms up and curl their fingers downward, pausing only to see which nail should be orally whittled next.

Do you have the hands of a concert pianist? Or a murderer? Or a concert pianist who becomes a murderer—remember that old Conrad Veidt horror movie, *The Hands of Orlac*, where he loses his hands in an accident and they sew on a murderer's hands and then Conrad gets all murdery?

Uh, where were we?

Looking at the length of the digits, specifically the second and fourth ones, women see equality (an issue they extend well beyond fingers), whereas men are more likely to have a longer ring finger than index finger. Some may say that this is God's constant reminder to hold out until your wedding night. But the real result of the fingers' disparity: A man is more likely to clench his fingers into a fist with a longer ring finger. This is called the Casanova pattern and is thought to result from exposure to androgens (male hormones, such as testosterone) in the womb. Preliminary studies have linked this trait (in women, as well) to shorter tempers and an analytical ability that makes for natural physicists, mathematicians, and the sort of scientist who might be drawn to comparing finger lengths. And just to be clear, the longer ring finger does not just mean talking tough or acting like a bully; it specifically refers to a higher incidence of fisticuffs and a greater need for anger management counseling.

However, before you start punching the clock at your punching-bag club, you should know that these findings are preliminary and that the biggest predictor of finger length is—you guessed it—heredity. No matter how long the ring finger, it's your family's 6-foot frame that will crush those dreams of becoming a jockey and maybe will make your kid mad enough to punch something. Now, if there's some speed and power in that punch. . . .

Bad Finger

Now that you've been looking at your fingers and deciphering what they mean, how about a little quiz for people who let their fingers do the talking?

American Sign Language users and deep-sea divers whose mouths are full of mouthpieces pay close attention to the hands of other people. Do you know what they're saying?

Gesture	Divers	Signers
Point sideways with both index and middle fingers (but hold your thumb down).	Look over there.	It's the letter *H*.
Thumbs up (as in "Let the gladiator live!").	Let's go up.	That's the number *10*.
Point with your index fingers on both hands; bring them together side by side.	Get with your buddy.	Together.

Gesture	Divers	Signers
Clasp hands together (don't intertwine fingers).	Let's hold hands	Marriage. (Clasped together just once. If you turn your hands and clasp them the other way you're making a burger.)
Cup both hands together in front of you.	That means *boat*.	That too means *boat* if you bounce your cupped hands a few times over the choppy waves.

Lap Dance

Mobile you, a laptop is your constant companion. You keep in contact with colleagues, monitor the market's every pulse, listen to streaming audio, and blog in real time. Terrific. **But for many users, there's one drawback to having a lap dance with your computer. What is it?**

Don't Get All Hot and Feathered

Men tend to keep their manly parts a good distance from certain things (i.e., knees, fast-moving objects, unpegged lobsters). In addition to following the usual doctor's orders (boxers not briefs, baggy jeans not bicycle shorts), men who look forward to fatherhood may use their laptops many places—airports, coffee shops, and all those other de facto branch offices—as long as they keep it away from one place in particular: the lap.

All those games of Minesweeper or solitaire aren't just killing time; the average laptop generates enough heat to kill healthy sperm, and a lower sperm count decreases the chances you'll ever have someone to blame for accidentally knocking your beverage into your keyboard. A switched-on laptop raises a man's scrotal temperatures by nearly 6°F, and men can experience a 4-degree increase even when the computer is turned off (no pun intended). (Oh, all right, let's go with the innuendo just this one time.)

To produce healthy sperm, the testicles must maintain a temperature that's lower than the rest of the body, which is why the testicles have their own sac external to the body. It's nature's way of regulating temperature in an uncertain world where temperatures fluctuate from the subtropical heat of a feverish influenza (during which testicles wisely descend to cool the sperm by getting as far away from your fluey self as possible) to the freezing New Year's Day dip in the sea with the Polar Bear Club (elevator *up*—next stop, nestled up next to some nice warm innards).

Low sperm count is the leading cause of infertility. And as men grow older, the count only decreases: Men younger than thirty-nine have fertilization rates of more than 60 percent, whereas those older than thirty-nine average only about 30 percent.

What else should men avoid to keep the sperm swimming and the potential for offspring afloat? Other sources of heat, of course: excessive use of saunas, hot tubs, or steam rooms. Smoking. Drugs. Stress. Oh, and your bike. That's right: Dedicated riders are exposed to pressure from the narrow seat that can bruise blood vessels needed to create and sustain erections. All those vibrations can damage the perineum (the scrotum's far underside), and these repeated microtraumas can decrease sperm counts.

Curiously, laptops are not yet among the contraceptives banned by the Vatican.

Heh-Heh-Heh, You Said "Bonus"

Ah, hanging out on the street corner, home of doo-wop groups, comic book traders, and sex education. Everyone has probably gathered a helpful hint or two there. Yet do any of these words of wisdom work for would-be fathers?

Which of these statements are really street wise?

Stress, tension, and anxiety can double a man's sperm count. No, all these factors help keep sperm production low. Not wise.

Drinking and smoking relax a man's nerves and increase fertility. Tobacco use damages sperm DNA, as does too much alcohol. Marijuana and other drugs can also lower sperm count. Time to go on the wagon (but not the bumpy, microtrauma-inducing wagon). Not wise.

Men who want to become fathers should put on a few pounds—say, 10 to 20 percent more weight than the national average. Excess body fat can cause all kinds of health problems, from poor circulation to dangerous hormone imbalances. Even a few extra pounds can lower your energy to "not tonight, dear" levels. Not wise.

Men who want to have children someday should have as many partners as they can as early as possible: Practice makes babies. The kind of dance (that is, the safe kind) matters much more than the number of dance partners. Increase your chance of exposure to sexually transmitted diseases *and* decrease your fertility. Not wise.

Men who lavish their partners with affection are more likely to become fathers than those who don't. Remember, first you've got to convince someone to go out with you. This one really *is* street wise.

Give My Regards to Magnetic Resonance Imaging?

If you're an aspiring opera singer or a Broadway musical actor, an MRI can do one thing for your career even more accurately than your vocal coach can. **What would that be?**

You and Your Big Mouth

It can make you a star! To ensure your onstage success early in your career, skip the silicon—or whatever they're singing about in *A Chorus Line* to "tighten up the derriere"—and quickly find out whether you have the equivalent of a four-lane highway vocal range. Whether you're a soprano, bass, or somewhere in between, skip the butt abutment and head straight for the giant magnetic scanner.

It's well known that vocal instructors sometimes make mistakes. For instance, *someone* let Eddie Murphy record "Put Your Mouth on Me." When a vocal coach misclassifies a student's vocal range—for example, calls a baritone a bass, a soprano a contralto, or a tomato a tomahto—the wrong training not only strains a singer's natural voice but also increases the chances for developing vocal cord nodes and hoarseness, bad news for nearly everyone except certain emo bands and Bob Dylan. Singing with emphysema is assuredly unpleasant; *sounding* like you're singing with emphysema is precisely what kept Louis Armstrong so lovable even when he gave the trumpet a rest.

If an instructor with years of pro experience can't make a foolproof assessment, what can an MRI do for the arts? The answer: measure the length of your vocal cords and the width of your vocal tract. (Curiously, tongue length determines only how likely you are to dress up in outlandish costumes and paint your face like Gene Simmons.) An Italian study has shown that longer cords, wider tracts, and bigger mouths mean lower voices; shorter cords and narrower tracts create higher ranges. This means Ezio Pinza's cords were probably twice the size of Rosa Ponselle's. Translated into popular music, that would mean Barry White had vocal cords like jumper cables and Dolly Parton does not. See, size does matter.

Unfortunately, even an MRI can't see whether you have "that special something" deep inside you. For that, you need someone to saw you in half.

You Can't Run from Your Fate

Take off your shoes and socks and look at your feet. **(Go ahead—we're all used to seeing odd behaviors in public.)** What professional potential might your toes reveal about you?

Shorter's Better, at Least in the Long Run

If your toes are long, you might want to reevaluate your plans to become an Olympic marathon runner. (Also persuasive: The moment you learn that Philippides, the first man to run a marathon, supposedly collapsed and died afterward.) Science has proven that short toes are superior for running and even for walking.

Short toes need fewer muscles to do the same work as long toes, and they need less motor force to stabilize the foot during each step. We pulled ahead of the great apes with our short toes, which made us more fit for running down food on the vast savannas and which make modern humans more fit for eating fast food on the run. The benefits of short toes include fewer injuries, better coordination, and less energy output needed for movement, all of which gave *Australopithecus* (the first bipedal hominid) a leg up.

Can your feet foretell more than Olympic prospects? Ostensibly, if your second toe is longer than your first, you're dependable, and if your third toe is longer, you have a temper. And if your toes are webbed, you're in the company of such famous syndactylics as Ashton Kutcher, Dan Aykroyd, and Marge Simpson. (Josef Stalin was said to have had joined toes on his left foot.) Your chances of being born with webbed digits increases by 29 percent if your mother smokes while pregnant.

Along with 52 bones (about a fourth of your entire skeleton), your feet have a total of 66 joints, 214 ligaments, and 38 muscles that enable you to take an average of 8,000 to 10,000 steps each day, or about 115,000 miles in a lifetime. That's enough to pace the entire width of the United States forty-two times, as if you were nervously waiting for your state to deliver a fiscally sound budget.

Tie One On

Attention thick-necked jocks, jarheads, hardbodies, and suits: The next time you knot your four-in-hand, you'll want to stay loose to avoid what?

A Perilous Apparel Answer

First of all, this has nothing to do with the always embarrassing worry about whether it's skinny or wide ties that are cool at the moment. (The only constants are that bowties should be confined to tuxedo rentals, clip-ons to fourth graders, and bolo ties to eccentric Texas millionaires.) Whatever style you choose, your neckwear shouldn't be eye-poppin' tight.

One study showed that, besides enhancing your double chin and turning some bosses of the yelling persuasion an unsavory shade of crimson (not a good color on anyone), your buttoned-down look can harm your health. A tightly knotted necktie can compress the jugular vein and cause vasoconstriction; this increases blood pressure in the eye, which can damage the optic nerve and lead to glaucoma. The same study also showed that wearing a tight tie during glaucoma tests can elevate your blood pressure and produce inaccurate results.

Now are you convinced that every day should be "casual Friday"? No? Well, how about this: Tight neckties can cause restricted breathing, headaches, and eye pain, and they predispose you to progressive optic nerve degeneration and eventual blindness.

While we're on the subject of what's good for you, maybe you should reconsider the very idea of wearing a necktie. They are veritable theme parks for germs; one study in a New York hospital found bacteria and pathogens on almost half of the neckties worn by doctors. For sanitary reasons, British hospitals have banned them altogether. If *that's* not proof that a tie is just about the most unsavory and dangerous item of clothing, why is it practically the first thing they confiscate when the nice judge reserves you a night in a jail cell on a drunk and disorderly charge (or so we hear)?

If you just can't kick the habit due to your ultraconservative leanings or your prep school's requirements or because your boss needs a stickectomy, go ahead and keep the tie. Choose skinny or wide, but whatever you do as you're knotting, hang loose.

Ode to Joy

On your way for a run, you accidentally grab your son's MP3 player and race out the door. **How did his playlist affect your workout?**

Pump Up the (Blood) Volume

Unless you're into Pink or Akon or Morrissey, his playlist made you feel tired more quickly than chasing him with a comb on school picture day. And if he'd grabbed *your* music, Barbra and Basia and Rod would have him winded in no time. Why so? Music that puts you in a good mood—as opposed to genres that make you reach for ibuprofen or drum your fingers with boredom—expands blood vessels and increases blood flow, improving your cardiovascular fitness. To achieve the opposite results, exercise to recordings of your neighbor's two-year-old at naptime or your spouse's morning shower serenades.

In an experiment, ten volunteers listened to songs they said made them anxious. They experienced a 6 percent contraction in their blood vessels. Then they listened to music they described as joyful. Their blood vessels increased in size by 26 percent. Participants chose different genres for their joyful tracks: Country fans might have found that Johnny Cash's "I Walk the Line" helped them *jog* the line, 50 Cent's "Candy Shop" could have pumped up a rap fan's heart, and almost anyone who chose "Gonna Fly Now" would have sprinted up the steps of the Philadelphia Museum of Art just like Rocky.

Listening to music releases endorphins, our natural stress fighters. The concept of music therapy dates back to the writings of Plato and Aristotle, when tortoiseshell lyre solos rocked amphitheaters all over the republic. (Crowd surfing was not invented until after the toga was replaced with something less inclined to expose the surfer.) Mothers in labor, children with learning disabilities, those suffering from Alzheimer's, and people with cancer are among those for whom music can provide comfort and physical, psychological, and emotional healing.

The upshot: Respect your biorhythms! Groove to what moves you.

Don't Lose Sleep over This One

It's the little things that annoy you most. **"We keep trying to throw things at them, but they are outwitting us,"** says Harvard's Richard J. Pollack, referring to something that has plagued humans for centuries. Who are they, and why are they stalking us?

The Creepy Crawly Clarification

They sound like the sarcastic guys in the IT department. Or, worse, the nonprofit groups that keep sending us free return address labels so we can save the world—or a portion of it, anyway—just by renewing a subscription to their newsletter.

Dr. Pollack may share those concerns, but he's a parasitic insect expert who's talking about guests unwelcome everywhere: bedbugs. You know, *Cimex lectularius,* the critters that sneak into your home like a tiny battalion of Santa Clauses to deliver a wealth of itchy, reddish welts as presents.

For these nocturnal bloodsuckers, every day is Christmas. And they're back, in a not-so-small way—merrily infesting bedding, headboards, floorboards, box springs, clocks, cracks, phones, curtains—turning up everywhere from run-down motels to lavish condos all across North America. Drawn by your warmth and carbon dioxide, they slit your skin, sip for five minutes, and stagger off to a nearby cranny to nap. In fact, urban entomologist Dr. Michael F. Potter warns that bedbugs are destined to become "the pest of the twenty-first century."

These bite-sized vampires have five life stages; by the time they have matured from egg to nymph to adult, they're nearly a quarter inch long. Bedbugs lay up to five eggs each day and feed every seven to ten days. Do the math: That's a lot of bedbugs. And like some overhyper-caffeinated collegiate procrastinator, they peak at about 3 A.M. It turns out that "Sleep tight, don't let the bedbugs bite" is actually a very sound goodnight wish.

Although their bites don't usually hurt and they aren't known to transmit human diseases, bedbugs are irritating enough to cause victims to trash mattresses, scrub their living areas with ammonia and bleach, vacuum their books, and even sue landlords for failing to exterminate these mother-suckers. What are the three most important words in bedbug real estate? You guessed it: Extermination, extermination, extermination.

Take Two Bedbugs and Call Me in the Morning

Bedbugs are known as the stinking bug to the Chinese. The Greeks named a spice coriander (*koros* = bug) because it smells like a crushed bedbug. Whether you are sweet on them or your relationship has soured, bedbugs have been our miniscule muses from time immemorial, friend to doctors, historians, playwrights, poets, and exterminators alike.

Pliny the Elder's *Natural History* credits the bedbug with several cures for what ails us, including night fever ("bind a couple to your left arm with some wool you've stolen from a shepherd"), malaria ("swallow seven bedbugs with a little meat and beans"), and epilepsy ("take bedbugs at the start of a seizure"). Other scientists agree: Bedbugs are as medicinal as they are adorable. Dioscorides favored bedbugs as a cure for malaria and prescribed them crushed into tortoise blood to heal wounds. Traditional healers in India have used bedbugs as a cure for hemorrhoids, urinary complaints, snake bites, and male pattern baldness.

Bedbugs make us itch, but they also move us to song: There's the "Mean Old Bedbug Blues," Shostakovich's Op. 19, Intermezzo (Suite from *The Bed-Bug*), and even Vladimir Mayakovsky's *The Bedbug,* in which our hero, Prisypkin, sings a song to entice an escaped bedbug to return to him—and what bug wouldn't? After all, don't we all want someone warm, breathing carbon dioxide, and full of blood?

Food for Thought

If you're like most people **(and for the sake of argument, let's say you are)**, you might guess that you do this about 15 times every day, but you actually do this closer to 200 times a day. What are you doing?

The Bowl-Is-Always-Half-Full Answer

Let's ask another question: Would you ever not finish a bowl half full of a soup you find delicious?

Probably not—or at least not soon. If the bowl is rigged to continually refill itself and to appear half full at all times, you'll eat and eat and eat.

This may sound like sci-fi technology for futuristic gluttons or some whacked-out experiment by a mad scientist; in fact, it's the latter. Dr. Brian Wansink of Cornell University's Food and Brand Lab cooks up unique tests to find out what we choose to eat and how much. And his experiments yielded some crazy results.

Wansink devised his bottomless bowl experiment to investigate whether people stop eating as a result of visual cues or because we actually feel full. Those who ate from a normal bowl consumed an average of 9 ounces, but those duped with endless soup ate an average of 15 ounces. Some participants consumed an entire quart (32 ounces, or about three cans of the condensed stuff), gorging for all twenty minutes of the experiment. (Editor's note: Sorry, no word yet about a bottomless beer helmet experiment.)

More proof that size does indeed matter, especially in making you fatter: Dr. Wansink conducted a study in which moviegoers received free stale popcorn, and those who ate from large containers ate 53 percent more than those with small boxes. Maybe not a surprising result when you consider these people are used to shelling out $12 for a salty movie-time snack.

As for what you do 200 times a day: You make choices about food—and not just when you're standing at the lunch buffet. Even people who aren't dedicated foodies make dozens of decisions at every meal (not to mention between meals) about eating, finding out what someone else is eating, turning down polite offers to share what that person is eating, changing your order when the burger arrives at the next table, snacking, dieting, cheating, fending off the "helpful" folks who offer to add fries, supersize, or "make it a combo for another $1.95."

Still feeling hungry?

Waking Up on the Right Side of the Bed

If you want to decrease your chances of being in a car accident, boost your income some 30 percent, take fewer sick days, and live about five years longer, **how should you begin each day?**

The Tongue-in-Someone-Else's-Cheek Answer

You may wake up with some "good to the last drip" coffee during your "breakfast of champions" and stop on your way to work to pick up a lottery ticket with your "dollar and a dream," but, come tomorrow, you won't be healthier, wealthier, or wiser (especially if you're buying lottery tickets, because you don't understand how statistics work).

The best way to improve your life is right under your nose: Kiss your sweetheart. Scientists in Germany have found that those who kiss in the morning begin the day with a positive attitude; the experience is even more positive if you don't forget to brush your teeth and floss.

Although philematology (the art and science of lip-locking) remains largely unstudied, researchers have proven that kissing not only decreases levels of cortisol, a chemical that creates stress, but also increases oxytocin, a chemical that heightens feelings of affection and potentially fights depression and disease. (One exception is lovesickness, a condition caused by watching your favorite restaurant become amateur night when it's filled with cooing couples on Valentine's Day.)

No matter what day it is, all dogs, some butterflies, and most people truly love kissing: 90 percent of us do it, and 85 percent say it's *the* most sensual activity. Whereas women tend to associate kissing with romantic love or long-term attachment, men often think of kissing as, let us say, a means to an end. This is one reason men tend not to wear lipstick and why they prefer their kisses wetter and sloppier, which also explains why canines, rather than women, are man's best friend.

Another positive result of kissing: It's a workout! A passionate smooch involves up to thirty-four facial muscles, increases your heart rate, and burns 6.4 calories per minute. Compare that with the 11.2 per minute you burn on a treadmill, training like an Olympic champion who may never be featured on a cereal box. Pucker up and live the good life!

Spin-the-Bottle

Supply answers to these three questions:

1. If you want to slow your body's aging process, you should ______________.

2. To prevent tooth decay, destroy plaque, and keep your gums lubricated and healthy, you should ______________.

3. Remember this passage from your college biology textbook? "For *homo sapiens,* pair bond reinforcement, which works to create the requisite reproductive capacity of a genetic pair and the viability of its progeny, is most readily expressed by a ____________." (Don't remember? Then guess! It's what you did back in college, anyway.)

Turn the page for the **answers**.

Answer

In each case, the answer is *kiss*. Kissing tones the jaws and cheeks to reduce the sagging of aging. Kissing produces saliva that improves dental health. And kissing is a boon to couples. It keeps them close (and, importantly, too busy to talk, which can only lead to arguments). A friendly peck on the cheek makes for a general feeling of goodwill and bonhomie (plus, it makes you seem *tres* continental). Smooches are an ongoing expression of affection that keeps families close throughout the terrible twos and the thuggish thirteens. And, as we all know, kisses can make almost any pain—from bumps and scrapes to second place at the science fair or the lacrosse meet—all better.

Will You Think of Me Sometimes, Doc?

Why is it that doctors think so often about Nathaniel Highmore, **Albert Wojciech Adamkiewicz, and Sir Rutherford Alcock,** and we don't even know those blokes?

He's Got My Sinuses

Doctors think of everything. They remember to bring in their old magazines for you. They go to school forever. They hobnob with tiresome drug salesmen. And they memorize thousands upon thousands of names for bones, muscles, and physiological processes named after people such as Mr. Highmore (the antrum of Highmore is the sinus behind your cheek), Mr. Adamkiewicz (the Adamkiewicz artery supplies blood to your lower spine), and even the benighted Sir Alcock (Alcock's canal accommodates the plumbing that operates a man's naughty bits).

Eponyms—no big deal, right? They're things named after their inventors and discoverers and the victors who wrote history. Just as there have been eponyms in geography such as Pike's Peak and Pennsylvania, so scientists have hardly been shy about naming guts and gizzards and other internal geography after themselves.

Most of the sticky bits named after scientists and medical pioneers are body parts we wouldn't even recognize; after all, even the most shameless "discoverer" would have a hard time claiming the knee or the nose with his or her own little personal flag. By the time most of these parts are even visible, we're out cold thanks to the arts of a talented anesthesiologist.

So the next time you think (thanks to the great vein of Galen, named by Galen, the ancient Greek physician) that you might have a pain in your Rotter's lymph nodes or your pores of Kohn or your angle of Louis, just remember you have Josef Rotter, Hans Kohn, or Antoine Louis to thank for giving your pain a name.

Of course, the *origin* of that pain is something much more personal. Perhaps it was something of an air traffic control problem amid your series of wine flights? Perhaps it was Aunt Marisa's jalapeño surprise?

The Hip Bone's Connected to the Basin?

Judging from the names (even in Latin), our extraordinary body consists of pretty ordinary parts. In the lofty name of science or the lowly name of nosiness, folks have been poking around the human anatomy since the dawn of time (when they were permitted by religion and law to rummage around in cadavers, that is). Although many named their discoveries after themselves, some parts so obviously resembled well-known objects that everyone figured you had to call 'em like you see 'em.

Guess where you would find the body parts named for these everyday objects:

- **a)** Basin
- **b)** Trees
- **c)** Arms or branches
- **d)** Tunic
- **e)** Spiders
- **f)** Courtyard

Turn the page for the **answers**.

Answers

Pelvis is from the Latin word for "**basin.**" *Elvis* is the American word for "The King."

Your nerves are dendrites, from the Greek word for "**tree,**" as in, "If a dendrite falls in the woods, would anyone feel it?"

Okay, your arm is an **arm** because it is a **branch** of your body, but your lungs and your arteries also have "arms," and their official medical terms (e.g., *bronchi*) originate in words for something that branches off.

No matter what you're wearing, you're always overdressed because there are **tunics** or membranes that cover organs such as your eyes, genitalia, and spleen. (But you're never fully dressed without a smile, at least according to *Annie*. Seems unrelated, but you're probably smiling right now just knowing your eyes, genitalia, and spleen have their own formalwear.)

If you're looking for **spiders,** look no further than your own veins, your brain, and anything that looked arachnoid or spiderlike to the Greeks.

An atrium, a Roman architectural term, is a **courtyard.** You have two atria in the upper chambers of your heart and one in your intestines, although none offer amenities such as little shampoos and free continental breakfast like they do at Courtyard by Marriott.

How Does the Color Wheel Make You Feel?

If you were competing at the Olympics—**say, in boxing, wrestling, or tae kwon do**—you could increase your chances of winning by a full 10 percent if you do one simple thing, which has nothing to do with your training. What is it?

The I-Red-That-Somewhere Answer

If you wanted an unfair advantage, you might toss a little sand into your opponent's eyes or fix the match with a bookie. But you're an Olympian! A true amateur! You'd never lower yourself to chicanery or steroids (not until you have that multi–million-dollar major league baseball contract, anyway).

Your best bet to influence your match without resorting to ingesting substances is to call dibs on the red boxing gloves, the red singlet, or the red belt. (True, having a black belt may also help.) That's right, a study at the 2004 Olympics revealed that competitors who dressed in red defeated their blue-clad opponents 60 percent of the time.

Why? Researchers say that red subconsciously connotes dominance. Leaders of capitalist countries during the Cold War would not necessarily agree.

In another study where participants performed different activities on a red, blue, or neutral computer screen, the red group proved more attuned to detail; they could remember words and proofread documents more accurately. Those using a blue screen proved more creative and imaginative—for example, inventing different uses for an object provided by researchers.

Another study used green (along with red and a neutral color), and researchers found out that subjects who were provided tests with a red cover sheet came out with lower scores, whereas those with green cover sheets had higher scores. The reason, they claim, is that red connotes danger, warnings, and excitement—hardly the calm, focused emotional state ideal for taking a test.

Subjects using a red and green cover sheet doodled more reindeer in the margins of their booklets. Merry Christmas to all, and may the best man win!

Red Is for Danger, and for Love, and for Dangerous Love

Many cultures have their own unique attitudes toward colors; for instance, in the United States, people will deal with a difficult customer "as long as his money is green," but green doesn't symbolize money in Singapore. Here are some other color associations in different cultures:

	To Adults in the U.S.	To Kids in the U.S.
Red	Stop signs Love and romance Emergency	Cherry sno-cone Pimples Teacher's corrections
Yellow	Cowardice Caution Possibly banana, lemon, or vanilla sheet cake	Rubber ducky Mac 'n' cheese Matches ("Don't play with matches!")
Blue	Sadness (feeling blue) Loyalty (true blue)	Overalls Raspberry juice bottles Snowball fight ("I forgot my gloves!")
Purple	Royalty	Friendly TV dinosaur Grandpa's feet
Green	Money Envy Environmentalists Nasty stuff	Money Nasty stuff
White	Purity Innocence Surrender (flags) Cosmetic dentistry	School nurse's uniform

	To Citizens of Other Cultures	To Nurses and Doctors
Red	China: luck India: purity Cherokee: success	Inflammation Intoxication Embarrassment Fever First aid
Yellow	Union of Myanmar: mourning Saudi Arabia: strength China: royalty	Jaundice Infection Bile and chyme Bruises
Blue	Iran: mourning India: the color of Krishna Cherokee: defeat	Vein Loss of blood flow Loss of oxygen Bruises
Purple	Thailand: mourning Brazil: death Ancient Rome: status	Your liver Other internal organs Bruises
Green	Greece (ancient): victory White Mountain Apache: south Ireland: Ireland	Gangrene Infection Bruises
White	Japan: mourning Navajo: dawn India: sadness	Poor circulation Eyes (sclera) White blood cells

The Power of Shower

When your partner is "not in the mood" but you are, a cold shower might be suggested as a means of dousing your desire. However, cold showers can also have a profound effect on another sort of mood.

What would that be?

The High-Pressure Answer

So you think numbing temperatures will cool your ardor? Extreme coldwater swimmers (nearly naked people in water that's nearly frozen) swear that the sport actually boosts libido. And now science, bless her heart, has shown that the very chilly water that turns your skin blue might also turn around your blue mood.

Icing down in the shower may stimulate the "blue spot" (no prurient double entendre here, people; we've moved on), also known as the *locus ceruleus,* which supplies most of the brain's noradrenaline, the depression-fighting chemical that surely deserves a nobler, more superhero-like name. What about *Norad the Uplifter?*

Furthermore, chilly water stimulates nerve endings that register cold; since these nerves are three to ten times denser than warm nerve endings, the burst of stimulation delivers a mild electroshock to the brain. Electric chairs, electrical currents, and electric bills can produce much greater threats to your well-being, but this jolt lets you administer a bit of home-made homeopathy, a small, brief exposure to a potentially harmful condition that might actually strengthen your body's ability to repair and heal itself. The theory behind homeopathy is that during mammalian evolution, harsh conditions—hunting in cold waters or choosing Africa as the cradle of civilization—actually improved humans' ability to recover and toughened immune systems against future assaults.

Sometime after humans evolved from apes (somewhere about the crack of the nineteenth century), German priests popularized cold water therapy. Of course, Chinese medicine had been prescribing cold water pick-me-ups for millennia, and now we're proposing it. Ask your doctor whether you should try your own shower therapy: one-to–three-minute showers, morning and evening, in 60–68°F water for at least four weeks. It could stabilize your blood pressure. It might reduce aches and pains. It will definitely help you use up all those shower products brought home from hotels. If you're an older adult, suffering from back problems, or someone who takes fifteen minutes to dip anything more than a toe into the water, you're hereby advised to see a doctor before beginning the therapy.

A Little Self-Medication

If your blood pressure is through the roof, **you have a short attention span**, and your allergies are flaring up worse than usual this spring, you might not be doing enough of what?

Humor Is the Best Medicine

We'll give you a hint: You also do this while watching a Mel Brooks movie, a YouTube video of a sneezing panda, or a sleeping person whose one hand you've dunked in warm water and whose other you've sunk in cold water. And although you don't have to pee yourself in the process, doctors recommend laughing more often.

Laughter releases natural antihistamines that work without prohibitive contraindications or the high price of over-the-counter drugs. Laughter also releases a mélange of beneficial hormones, cells, and enzymes in the body, including natural antihistamines—no prescription or litany of potential side effects needed. In one study, subjects were exposed to allergens that triggered their allergy symptoms (for example, kids and peanuts, adults and pollen, senators and honesty). After half were shown a ninety-minute comedy video and then exposed to their particular allergens again, their allergic reactions were significantly milder, and the positive effects lasted for over four hours. (Results on the senators were harder to determine: They fake-laughed throughout the entire film and continued to lie afterwards.)

Biochemically (doesn't that word look handsome in front of a comma?), laughter deploys hormones and enzymes in your body that fight disease and contribute to your overall feeling of well-being. As the cartoon squirrel waits behind the tree with a baseball bat, your body activates NK cells (lymphocytes known as "natural killers"). When the dog runs around the tree, the cytotoxins go to work, destroying tumors and viruses. Stars fly from behind the tree, and your immune system releases gamma-interferon to bring disease to its knees.

Laughs (real hilarious laughs, not the-boss-thinks-he's-funny kind) actually decrease our stress hormones and boost our immune system and attentiveness. Aside from working against sneezing and fidgeting, laughter seems to promote growth of the endothelium, the protective lining of blood vessels, which decreases cholesterol build-up in coronary arteries and protects against heart disease, the leading cause of death in the United States.

Laughter is also an expression of mental health as long as it's socially appropriate (e.g., no pointing and laughing at the new chainsaw owner who can't even get it started; he just won't get it). Whether the news is good or bad, there are times in life when cracking up keeps us from cracking up.

You think this book is funny, right?

Right?

See? It's helping!

Just for Kicks and Giggles

Sure, yukking it up can boost your immunity, elevate your mood, and banish stress, but laughing can also mean other things depending on the context and your state of mind. It can indicate happiness, surprise, a desire to seem attractive or appealing, nervousness, ticklishness, agreement or the willingness to go along or get along, derision (as when holding someone up to ridicule), an attempt to confirm another's vulnerability (think supervillain), or simply a way to let the world know you are under the influence of laughing gas or another substance (okay, let's just stick with laughing gas for now).

Okay, so laughter isn't always pleasant, but perhaps it will make you feel better to know that someone who laughs *at* you is not getting all the health benefits enjoyed by someone who laughs *with* you. Moreover, laughter, like smiling, can be forged, which was one source of angst for the original crazy, mixed-up kid Hamlet, who gnashed, "That one may smile and smile and be a villain."

Despite its health benefits, laughter is not recommended in situations when misfortune befalls the powerful, especially if those in charge lack humility sufficient to laugh at themselves.

For animals (and some humans), smiling isn't so much an expression of joy as a show of submission or aggression—a threat behavior known as baring the teeth. Although they don't seem to laugh much, recent developments in technology allow us to see that babies smile in the womb. Pessimists will say that's because they are not yet familiar with diaper rash, standardized testing, or trying to book a flight using award miles.

Some studies in gelotology (yes, it does sound like the study of Jell-O and its jiggly brethren) suggest there are differences in the ways men and women use, express, and appreciate humor and laughter. However, nobody is funding that sort of research these days.

A Question About Test Questions

You're a high school teenager prepping for college admission tests. **You'd do anything to be successful, but you can only study so much.** Scientists have now proven that there is one more thing you can do to give yourself an extra edge. Twice a week, you should do what to boost your scores?

The Fish-Food-for-Thought Answer

When it comes to prepping for the SAT, ACT, and various other acronyms that are supposed to guarantee Success with a capital *S,* nothing guarantees a great score. You tried everything to increase your scores, but the test prep courses degenerated into expensive naps, tutoring hours turned into expensive make-out sessions (trigonometry *is* nature's strongest aphrodisiac), and eating salmon for dinner twice a week improved your score by over 10 percent.

You heard correctly. According to a Swedish study of nearly 4,000 teenage boys, those who ate fish once a week increased verbal intelligence scores by 4 percent and visual-spatial scores by 9 percent. Students who ate fish more than once a week increased their scores by 9 and 11 percent, respectively. (Surely those who ate fish at every meal are the ones writing college admissions tests.)

Aside from studies about fish and brains and fish brains (which you'd think would be a lot bigger, considering they're in school all day), researchers concluded that since fish are high in omega-3 and omega-6 fatty acids (the acids themselves prefer the term *voluptuous*), this helps with intellectual development. Besides the good news about fish and brain power, studies have shown that omega-3s help decrease a woman's chances of early delivery and low-birthweight infants; they're also good for the baby's development.

And fish is good for grown-ups: Those of us who eat fish are 36 percent less likely to die from heart disease than those of us who don't. So have a (smart) heart: Eat more fish.

Hirsute Yourself

Men create this more often than women. **Older men create it more than younger men.** Other mammals don't create this at all. What is it, and can you sell it on eBay?

The Omphaloskeptical Answer

All of us are navel gazers at some point in life because everyone sports a bit of belly button lint. This conglomeration is typically blue-gray, the sum color of all the clothes you wear, despite the fact that using all the colors while finger painting results in something muddier. Science can explain this mystery.

A chemist named Georg Steinhauser conducted risky experiments for the good of all humanity: He saved more than 500 days' worth of lint. His discovery: Such masses are composed of dead skin, fat, dust, and clothing fiber. Human hair is a scale-like structure that acts like barbs to capture strands of fabric and sloughs them toward the ends of the hairs, straight into the old lint trap. (Women's body hair tends to be thinner and shorter, so it accumulates less lint; older men's bellies boast the fuzziest navels, and younger women drink the most fuzzy navels.)

Other researchers who laid the groundwork for the current state of our lint knowledge include Graham Barker (who collects his lint in jars and currently holds the Guinness world record for this pursuit) and scientist Karl Kruszelnicki, the theorist who took the belly button world by storm with his we-wear-gray-clothes-so-our-lint-is-gray theory. Perhaps one of these brave Australians (weird how they're all Australians, right?) will be known one day as "the father of modern belly button fuzz."

The Ins and Outs of Navels

There are no health risks to possessing navel lint, so cross that off your list of things to worry your doctor about. On the other hand, there is the subject of umbilicoplasty, a novel bit of cosmetic navel surgery.

Ever asked someone whether they have an Audi, only to hear them reply, "No, an innie!" Whereas 90 percent of us do possess an innie, 10 percent of the population would answer no, and an unknown proportion would go on a tangent about high-performance German automobiles.

Sure, some might be self-conscious about their navels, but at least after the umbilical cord is detached and ceremonially buried in the backyard (well, we know at least one family who does that), the stump that's left over dries up and falls off.

So why the abdominal apartheid? Why aren't we all either innies or outies?

Some doctors say that it's completely random, and others claim it's determined by how the cord is cut. A third theory is that outies are a result of umbilical hernias, a small circular growth found in the abdominal region of 10 to 30 percent of newborns.

Oddly, there's just no consensus in the medical field. Why? Because the only *real* authority, at least according to any four-year-old we've asked, is the stork, silly!

Don't Cell Yourself Short

Although we humans like to take pride in our individualism and personal achievements, **in reality we hardly accomplish a single thing alone**. Who really deserves the praise?

It Takes Guts

When you call someone else an inhuman name (a rat, for instance, or a dirty dog or an absolute pig), you don't mean it literally—that is, unless you point at someone and yell, "You, you bacteria!"

Our bodies possess 100 trillion bacteria cells, and that's ten times more than our human cells. Try as you might to rid yourself of them by slathering on disinfectant, wearing gloves, or refusing to give your friend a sip of your $12 martini, you can't possibly oust them all. Anyway, you wouldn't want to: Many types of bacteria are good for you.

In studies of rodents, a germ-free mouse had to consume three times the calories of a normal mouse to maintain body weight; later, when given a dose of bacteria, the sterile mouse gained significant weight despite an average diet. Bacteria act as catalysts to break down food so we can derive energy and nutrition from the junk we shove in our pie holes (hurrah!), although some sugar-loving bacteria also adhere to your teeth and cause cavities (boo!)

Dozens of products fortified with probiotics (helpful bacteria) are vying for space on retail shelves and in your gut. The jury's still out on many of the manufacturers' claims, so ask your doctor before you fork over your dough for products that may or may not help your irritable bowel syndrome, irregularity, or other troubles. But one thing is certain: Bacteria occurring naturally in your digestive tract are the unsung, behind-the-scenes heroes that bolster your immunity, keep you well nourished, and fight infection. Let's raise a glass of drinkable yogurt (now with more varieties of acidophilus!) to the little guys, all 100 trillion of them.

Do You Give It the Nod?

In a given year, half of us admit to fighting the urge, but more than a quarter of us actually give in to it—**with dire consequences**. And the solution is literally at hand. So, please, let's wake up and put a stop to it. What is it?

Pull Over for the Answer

Yes, the temptation to multitask behind the wheel is almost irresistible, and we all have stories of drivers who endanger life and limb (ours) as they chat on the phone or review the performance of their fellow drivers (thumbs down, finger up). We all know it's dangerous to buckle up after a few beers, but how many of us hand the keys to a designated driver when we're feeling sleepy? Falling asleep is merely tacky at your niece's band recital (please, tell us: how do you drown out the squealing violas?), and it won't score you points at jury duty, but it's just plain risky to snooze behind a steering wheel.

However, this generation of American adults gets 20 percent less sleep than their parents. Some of it is voluntary (staying up late to sew more pearls of silliness on the conversation string on your Facebook page), and the rest results from the demands of work, family, commuting, and underlying conditions (such as obesity) that make us toss and turn. About 250,000 people a day doze off while driving. Not only do 28 percent of us actually fall asleep, more than 54 percent acknowledge they feel drowsy on the road.

Drivers overcome by sleep are responsible for an estimated 1,500 to 5,500 deaths and 40,000 to 71,000 injuries per year. What's more, our country is just beginning to realize that sleepy drivers can be as dangerous as drunk ones. In fact, going eighteen hours without sleep impairs your driving performance as much as a blood-alcohol content of 0.08 percent. Depending on your weight and what you've eaten, that's the equivalent of downing two, three, or four rum-and-Cokes—well over the legal limit.

Driving while fatigued ensures a nightmare for you and anyone else on the road.

Soft shoulders. Soft, cuddly, warm shoulders . . . huh, WHAT?

The Soporific

Do you earn your living by racking up the miles, whether touring half-conscious in a family folk band, cruising the neighborhood in an ice cream truck, or hauling hazardous cargo cross-country? Are you one of the 2.4–3.9 million licensed big riggers with obstructive sleep apnea, a condition that disrupts nighttime sleep and makes for less alertness the next day at the wheel?

One suggestion for you drowsy road warriors: Pull over, down 8 ounces of coffee, and nap for thirty minutes; that way, you get to cop some *Z*s and then rouse just as the caffeine kicks in.

Another suggestion: Let your sixteen-year-old drive for a while and see how quickly adrenaline can work its rousing magic on you.

Lardy, Lardy, Sweet Lardy!

It's snowing. You decide to shovel your driveway and sidewalk—you know, for a little winter exercise—but what if you shoveled without a coat? Is there any reason to shiver while you shovel **(and muffle your snuffle and shuffle through the snowfall and stifle your sniffle)**?

The Cold, Hard Fats

Whether it's popping dubious diet pills, nibbling at the largest section on the food pyramid, or strapping on a sauna belt, people will do anything to stop jiggling—that is, as long as they don't have to do anything at all. Now there's good news for all of us wanna-be lean, mean, non–green-bean-eating machines! There's a fat that actually burns calories, and chances are, you already have it.

Humans have two types of fat. The bad kind is white fat (think lard, cupcake frosting, or mayonnaise), which stores up calories and makes us plump. The good kind, brown fat, naturally burns calories when it's activated by cold temperatures, as seen repeatedly in tests with mice. In one study of obesity-prone mice on a high-fat diet, the rotund rodents lost 47 percent of their body fat after a week in 41-degree temperatures. (Makes you think, "Next winter, I'm hitting the treadmill we've stored in the garage—where it's cold!" until you think, "Yeah, except that mice have probably made a nest of its circuitry by now.")

It was long believed that infants needed brown fat to stay warm until they could shiver effectively (the body's warming strategy), but new advanced scanners show that most adult humans also have little blobs of this lovable lard on their upper backs and necks—enough amazing adipose tissue to burn significant calories.

Leaner folks have more brown fat, which is named for the iron-filled, energy-burning mitochondria that tint the cells' tissues a reddish brown. Women tend to have more brown fat than men (score one for the girls) but also more white fat (score is tied again). Older people have less brown fat than younguns, even though the kids these days don't even walk ten miles to school every day in the snow. Although one Finnish source claims brown fat helps us shed up to 9 pounds per year, most studies—especially those in less frigid climates—confirm that merely shivering shirtless isn't an effective way to slim down.

Those infomercials may still be your best bet for doing nothing, but if you watch them with the thermostat turned way down you'll at least have something to show for doing nothing.

Hip Fashion

Let's say you're a very specialized clothing designer and your client requests an ultra-lightweight garment for work that will protect her **skull, sternum, shoulders, spine, hips, and thighs**. Who's your client, and what's with the strange demands?

Let's Marrow Down the Choices

With your client needing coverage in so many seemingly arbitrary places, you're tempted to just drop her into wet cement or sandwich her between two mattresses with duct tape (did you do an internship with the mob or something?).

What if we told you her workplace is at the end of a looooong commute?

That's right: Your client works in outer space, and although a space suit isn't supposed to be stylish (white just isn't slimming, and helmet hair isn't flattering), they do more than provide oxygen and make lines to the restroom a thing of the past. They also protect all those vulnerable bones mentioned earlier.

What's so vital about those five areas is that they house your bone marrow, which needs to be shielded from solar storms. When a star explodes (not a Hollywood star, although those also are famously volatile) the radiation released can wipe out an astronaut's bone marrow cells. That would not be a good thing, since bone marrow is what replenishes the body's blood cells. Without the bones' marrow, within a week not only would NASA piss off every vampire in the vicinity, but also the astronauts' immune systems would be compromised. Achy and exhausted, how could they tend those little gardens that always seem to spring up on space shuttles?

Blood-Sucking Bonus

How much of your body's weight is composed of vampire juice?

Okay, would-be astronauts, let's space-walk through this slowly:

Your largest bones play host to stem cells, and those differentiate into red or white blood cells, your two most important kinds. Red blood cells are responsible for conveying oxygen throughout your body, and they outnumber their white, disease-fighting sidekicks 700 to 1. The lipstick counter is the only other place where reds outnumber whites consistently.

Blood accounts for 7 to 8 percent of your total body weight, and red blood cells make up 40 to 45 percent of that. We've done the math: If you're a 150-pound person, you contain about 5 pounds of red blood cells. If you're 250 pounds, well, forget about being an astronaut.

Anyway, the important thing to take away from this is to wear your space suit no matter if it's unflattering and there are no restrooms or blood suckers in space.

Clothing Optional

Let's say you're sitting on a wooden box in the desert. Would you be more comfortable dressed in:

a) Standard olive drab army fatigues?
b) A light tan summer outfit?
c) As little clothing as your natural modesty and local law allow?

The Hot-Under-the-Collar Answer

All right, you're under the hot sun on the hot sand, so you're not comfortable. Unless you're a soldier during World War II and your other choices are chilling (so to speak) in the middle of the Sahara or dodging machine gun fire in Manila.

A few hours on a box may seem like no sweat. In fact, the volunteers in one study *did* sweat a lot, especially the near-naked ones, about 30 percent more than their clothed and fatigued (and therefore less fatigued) counterparts.

This experiment shows just how much heat your unprotected skin absorbs: Clothing keeps your body cool. In temperate weather, when you're unlikely to be sweating, excess metabolic heat seeps out of the skin and into the air, and some evaporation of water from skin cells provides further cooling. But when you start exercising or get caught in rising temperatures, there is too much heat and your skin stops losing it and absorbs it instead, thus elevating your body temperature. Next thing you know, it's ready, sweat, go!

Our apish ancestors, their bodies entirely covered with hair, couldn't cool themselves with sweat, but we smoother-skinned humans have 2 to 4 million sweat glands (and heavy sweaters can have up to five times more). Our temperature regulatory systems (perspiration, air conditioning, fanning the playbill in front of your face in a packed theater) make humans adaptable and able to survive in almost any climate.

Cooling off is even more critical than warming up. Hypothermia is serious trouble, but humans can survive a 20-degree drop in body temperature. However, a mere 6-degree fever fries the brain, and there's no surviving a fever of 110°F.

One city in America that benefits from those mutated sweat glands is Phoenix, Arizona, which Old Spice has proclaimed "The Sweatiest City in America" four out of the last six years. The cologne's parent company, Procter & Gamble, calculates that the average Phoenix resident sweats 26.3 ounces per hour, enough to fill 53,000 beer kegs in one hour, or an Olympic-sized swimming pool (660,253 gallons) in three hours. With all that perspiration (or "glowing" if you're a woman, or "foaming" if you're a horse), some Arizona residents might very well prefer Manila.

Now Pay Attention

Whether you're a tennis coach, a movie buff, a palm reader, an aspiring poet, or someone who does all four things at once, **you only get about 173 billion of these in your entire life**. Would you want fewer? More?

A Bit of an Answer

So you can read someone's lifeline with one hand and hit a backhand with the other while watching *Annie Hall* in your home theater and composing a sestina. Besides this scenario being completely outlandish—a poet couldn't afford a home theater—multitasking is a myth, according to author Winifred Gallagher.

When it comes to processing your lifetime allotment of 173 billion bits of information—a number that you can use up quickly if you watch really hard game shows—your brain's attention is constantly sought by a variety of competing stimuli that can truly exist in only one place at a time. (Are you listening, you daydreamers in the back row?)

Our brains focus on whatever is threatening, surprising, bright, loud, shiny—things behaviorists categorize as inherently salient, meaning we need to know about them. Also, we override that involuntary impulse with biased competition, which automatically makes our interests a priority. For example, you're at a swim meet, trying to locate your husband. The salient stimuli might be the sound of the coach's whistle, the shouts of the other spouses, and the smell of chlorine, but because of your attentional bias, your focus is on bringing your mate the pre-swim carb booster he forgot. Only after you carry out this vital mission does your brain remind you to appreciate the pool deck of muscular bods in wet bathing suits.

MIT researchers found that brains can't begin to process all the data the senses bring in. So signals compete for neural resources. In other words, your processing space is limited. Not only must neurons become active to focus on something, but they must synchronize, like swimmers—the synchronized kind of swimmers who can jump and splash and pretend to be sharks in unison. Consider how many other salient stimuli your neurons must ignore at any given minute of the day and night (your feet on the hot concrete deck, the rumble of your stomach anticipating lunch, that burning smell from the snack bar, your no-longer-minty chewing gum, the maintenance guy with his leaf blower) and you'll see why pickpockets work in crowds where people are jostling each other and why the testimony of crime witnesses can be so unreliable, even when Speedos aren't on the scene.

Mind over What's the Matter

While writing out a prescription, what information should your doctor make certain to withhold from you?

The License-to-Ill Answer

Could be the cost of the meds. Could be that he's a Dr. Kevorkian wanna-be. But the truth is, if your doctor describes the long list of a medicine's possible side effects, you're likely to have whatever sickness you're being treated for, along with a sidecar of other sicknesses such as nausea, headaches, ED, or any number of other potential ughs, blahs, and aches.

You've probably heard of the placebo effect—a harmless and inert substance can have a curative effect on patients who expect it to help them—but have you heard of the nocebo effect? Expectations can work both ways, and you probably *will* feel ill if you *expect* to do so. For example, in an early study, when patients were given a solution of sugar water and told it caused vomiting in some individuals, 80 percent of them up and upchucked. So if your doctor tells you a medicine might come with a laundry list of other troubles (such as nausea, headaches, erectile dysfunction), there's a chance that your subconscious will treat those symptoms as a to-do list. The good news is that half of American doctors regularly prescribe placebos to help difficult patients' psyches, a strategy that cures many perceived ailments.

In a multidecade study, women who feared they were at risk for a heart attack were 3.7 times as likely to die as a result of coronary problems than women who didn't feel at risk, even if they did not smoke or have other risk factors.

So keep belting your best Gloria Gaynor "I Will Survive" like you really mean it, and you just might.

(If You Feel Up to It)

It seems that humans throughout history have proven they will eat, drink, or apply almost anything to cure what ails them. Which of these remedies are a placebo, a nocebo, or a cure?

The Treatment	Cure or Quackery?
Greek and Roman soldiers spread honey on themselves after battle.	Sure, they liked to have a good time, just like anyone, but the soldiers weren't in the baths—they were in the infirmary curing spear wounds. Honey fights infection. Cure.
Old wives in the United Kingdom made mashed figwort.	Sounds like Scrooge's dinner, but this, too, can be applied to wounds to speed healing. Although modern scientists aren't sure why it's so effective (and why it must be fresh), it is a cure.

The Treatment	Cure or Quackery?
Ancient Egyptians recommend you fill your nose with lint and secure it from the outside.	It's not bad enough to break your nose; now you have to walk around looking like your schnozzola lost a fight with a taxidermist. Funny part is, this is roughly what we do today. Cure.

The Treatment	Cure or Quackery?
Your sister downs 3 ounces of cranberry juice every day, but she says it's not just for the vitamin C.	Pity your poor sister, who must have a urinary tract infection and is fighting its return via cranberry (actually *Lactobacillus,* a probiotic). If the UTI's not too serious, this is indeed a cure.
Medieval doctors in Persia prescribed garlic, frankincense, and rose oil for the same condition.	These odorous oddities treat headaches: fighting the cause of headaches in your body, relieving pain, putting you to sleep, or acting on the nervous system. So they're all cures.

Weighs Heavy on My Mind

What do plastic chewing gum wrappers, **cow belches,** and people who keep adding another hole to their belts all have in common?

The "Weight, Don't Tell Me!" Answer

Of all the creatures on Earth, human beings alone can take credit for changing the climate of the planet. By "changing" we mean "causing forests to become deserts," and by "climate" we mean "the weather on the planet's surface that makes all life possible."

Scoff if you'd like at the dire predictions of those who measure cattle burps, which contain methane, which breaks down the ozone layer that keeps our planet from frying. But can't we all agree that our landfills are choked with plastic and other materials that won't break down into mulch? That our love of cheeseburgers has driven the systematic destruction of forests for grazing lands? That the 10 pounds you gained since the prom helped to heap an additional 3.8 million tons of carbon dioxide into the atmosphere?

If you thought the tragedy was limited to the price of taking out your tux a couple of inches, you missed the Center for Disease Control's latest finding: Because of the average weight gain from the 1990s to 2006, airlines expend an extra 350 million gallons of fuel per year. (Of course, they could help cut fuel consumption if they cut consumption of calorie-packed peanuts and mini Crown Royal bottles. What if they started charging as much for saddlebags as they do for other bags over the weight limit?) Yes, our increasing obesity contributes to global warming: Lard-laden locomotion burns about a billion extra gallons per year.

Obesity has tripled since 1980, and our sedentary habits contribute to the rise of high blood pressure, heart disease, and Type 2 diabetes—in kids and adults. In 2004, the average American spent about eight and a half hours a day sitting and being entertained—a strange statistic considering that most Americans work, go to school, and, according to some sources, sleep.

Whether you carry around an extra hundred or merely ten pounds, they have an impact on both your health and the globe's. If we don't get off our butts, for the first time in our country's history, this generation of children could have a shorter life span than their parents.

So get up! Go out! Chase a dream! And while you're at it, work off that sugary prom punch (and the high-calorie booze you spiked it with).

You're Worth It

A New Year's Eve kiss: priceless. A Thanksgiving feast with your extended family: priceless. **A year of life: $129,090**. What's with that?

The Pieces-and-Parts Answer

If asked how much a year of your life is worth, you may consider several different factors. Am I likely to cure cancer, discover a new species of butterfly, or win *American Idol*? Does donating money to charity make me more valuable? What is the annual rate of inflation?

Wonder no more: According to a study in *Value in Health* (that's right, "the Official Journal of the International Society for Pharmacoeconomics and Outcomes Research"), a year of human life is worth $129,090. Of course, there are outliers—people who are only children, children from Orange County, dogs that billionaire widows treat (and endow) like children—but that figure is the study's "quality-adjusted" average. Specifically, that number is based on the cost of renal dialysis, a method of maintaining the chemical balance of blood after kidney failure, which is expensive but unconditionally covered by Medicare.

Researchers say the figure can "inform decisions such as the determination of compensation when life is lost to accidents or mishaps, and pricing and coverage of medical procedures and the pursuit of expensive but life-saving public programs"—not to mention whether there will be money left for private school, summer camp, or space tourism. Although every life is priceless, it turns out that some lives are more priceless than others. For instance, if a sanitation engineer and a corporate lawyer die in the same freak detergent truck accident, it's the corporate lawyer's kids who are going to clean up: Under U.S. law, when calculating things like damages, you are allowed to factor in things like net worth and annual salary and whether you already live in a mansion.

Celebrity

And then there are celebrities, stars who insure their body parts for more money than your life is worth. Many of these folks may have been ripped off, but at least they'd be compensated if their limbs were. Along with that of male stripper Frankie Jakeman (who took out penis insurance for $1.6 million), most of these policies would compensate for harm befalling people who are, let's face it, overcompensating:

Ben Turpin's crossed eyes: $25,000. The silent film star took out a policy in case his eyes suddenly and miraculously uncrossed. News of his insurance, the falls he took without breaking bones, his work as a custodian in the upscale L.A. apartments he owned, and his boasts of how he made $3,000 a week—all this secured Turpin's legacy as "that funny-looking guy who's not Charlie Chaplin."

Harvey Lowe's hands: $150,000. In 1934, the Cheerie Yo-Yo company created a sensation by insuring the magnificent mitts of the winner of their first World Yo-Yo contest. If harm should come to Harvey Lowe's hands, the company would come into an amount more than 100 times the average annual salary of those lucky enough to have a job.

Bruce Springsteen's voice: $6 million. Although it already sounds a bit scratchy, the Boss wants to make sure it's just roughed-up enough to supply the soundtrack for all those future Independence Day fireworks.

Tom Jones's chest hair: $7 million. Theft? Flood? Body waxing trouble? It's hard to guess how Mr. Jones could have grounds for a claim. But then again, insurance isn't for the foreseen but for the unforeseen. (Not that we wanted to even see the chest hair in the first place.)

Gennaro Pelliccia's tongue: about $16 million. The official taster for Costa Coffee said, "In my profession, my taste buds and sensory skills . . . and 18 years of experience enable me to distinguish between thousands of flavours." Odds are Pelliccia now has Earth's most expensive coffee breath.

David Beckham's legs: $70 million. What would the international soccer star be without his legs? Right: *still* the world's most successful model.

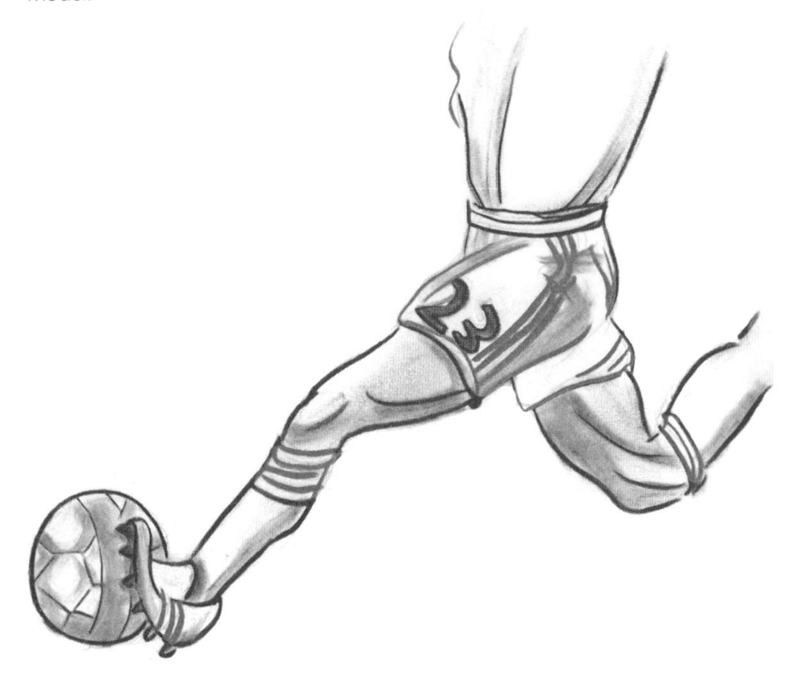

Appendix

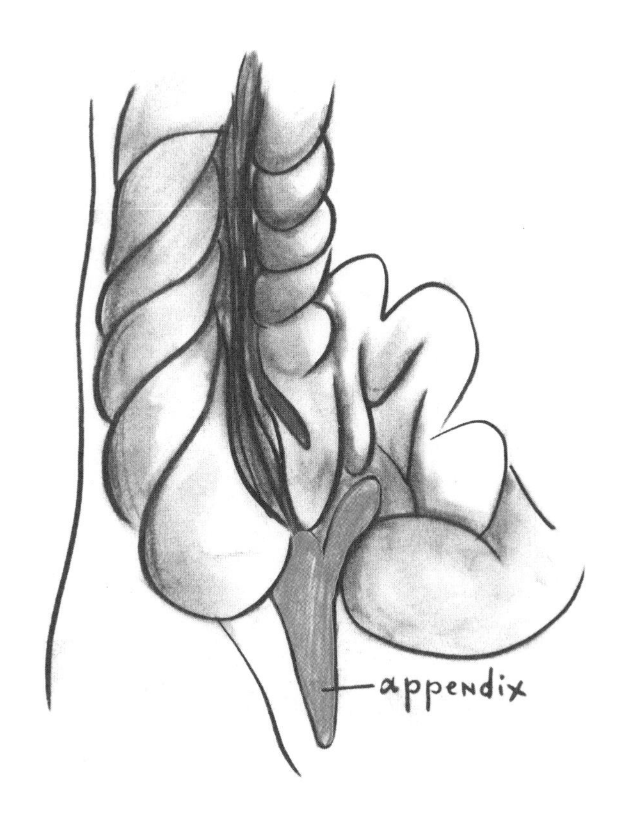

About the Authors

With his sequence of toy doctor bags, needleless syringes, prescription bottles refilled with candy pills, and a *genuine* stethoscope (compliments of Doctor Lil, his cousin and pediatrician), **Michael J. Rosen** spent ages four to twenty-six in preparation for becoming a doctor. And yet, after a semester at a Caribbean medical school, he decided to promote his writing and drawing hobbies to the status of a career. (Look, there are more things in life than getting drug-company-logo freebies and financial security.) Since then, he has written, edited, or illustrated more than eighty books for both adults and kids, none of which—until now—utilize the ~~pointless~~ countless hours he spent in three-hour, twice-a-week organic chemistry and physics labs. He lives on a hundred-acre farm in the Appalachian foothills of Ohio. His Web site is www.fidosopher.com.

Ben Kassoy, resident nonresident, placebo junkie, and Anti-brain Freeze Committee Chair, runs the Paper or Plastic Surgeon Volunteer Day at area grocery stores. With Michael J. Rosen, he authored a collection of stories about the world's most eccentric sports, *No Dribbling the Squid* (Andrews McMeel Publishing, LLC, 2009). He hails from Bexley, Ohio, and attends Emory University.

Writer and performer **M. Sweeney Lawless** was a founder of the comedy troupes Euphobia and The Central Fungus, and is a member of the improv theater Chicago City Limit's national touring company. Her work has appeared on CNN Headline News, the Huffington Post, McSweeney's Internet Tendency, *Lowbrow Reader*, some humor anthologies, and, most recently, in a book about the human body (*eeeuww*). She lives on a small island off the coast of the United States. Meg can hide up to five bees secretly in her mouth.